POLITICS, PRICES, AND PETROLEUM

POLITICS, PRICES, AND PETROLEUM

The Political Economy of Energy

By **DAVID GLASNER**

Foreword by
PAUL W. MacAVOY

Introduction by
BENJAMIN ZYCHER

Pacific Studies in Public Policy

PACIFIC INSTITUTE FOR PUBLIC POLICY RESEARCH
San Francisco, California

Ballinger Publishing Company
Cambridge, Massachusetts
A Subsidiary of Harper & Row, Publishers, Inc.

International Standard Book Number: 0-88410-953-4 (CL)
0-88410-954-2 (PB)

Library of Congress Catalog Card Number: 84-19134

Printed in the United States of America

Library of Congress Cataloging in Publication Data

Glasner, David.
Politics, prices, and petroleum.

(Pacific studies in public policy)
Bibliography: p.
Includes index.
1. Gasoline industry—Government policy—United States.
2. Petroleum industry and trade—Government policy—United States. 3. Gas industry—Government policy—United States.
4. Energy policy—United States. I. Title. II. Series.
HD9579.G5U5435 1984 338.2'7282'0973 84-19134
ISBN 0-88410-953-4
ISBN 0-88410-954-2 (pbk.)

With a prayer that both he and his wife, Eta, may enjoy many more years of good health and happiness together, I should like to dedicate this book to a great and good man, and a marvelous scholar from whom I have learned so much, Rabbi David Shapiro.

The Pacific Institute for Public Policy Research is an independent, tax-exempt research and educational organization. The Institute's program is designed to broaden public understanding of the nature and effects of market processes and government policy.

With the bureaucratization and politicization of modern society, scholars, business and civic leaders, the media, policymakers, and the general public have too often been isolated from meaningful solutions to critical public issues. To facilitate a more active and enlightened discussion of such issues, the Pacific Institute sponsors in-depth studies into the nature of and possible solutions to major social, economic, and environmental problems. Undertaken regardless of the sanctity of any particular government program, or the customs, prejudices, or temper of the times, the Institute's studies aim to ensure that alternative approaches to currently problematic policy areas are fully evaluated, the best remedies discovered, and these findings made widely available. The results of this work are published as books and monographs, and form the basis for numerous conference and media programs.

Through this program of research and commentary, the Institute seeks to evaluate the premises and consequences of government policy, and provide the foundations necessary for constructive policy reform.

PACIFIC STUDIES IN PUBLIC POLICY (*SELECTED TITLES*)

Money in Crisis
The Federal Reserve, the Economy, and Monetary Reform
Edited by Barry N. Siegel
With a Foreword by Leland B. Yeager

Forestlands
Public and Private
Edited by Robert T. Deacon and M. Bruce Johnson
With a Foreword by B. Delworth Gardner

Urban Transit
The Private Challenge to Public Transportation
Edited by Charles A. Lave
With a Foreword by John Meyer

Firearms and Violence
Issues of Public Policy
Edited by Don B. Kates, Jr.
With a Foreword by John Kaplan

Fugitive Industry
The Economics and Politics of Deindustrialization
By Richard B. McKenzie
With a Foreword by Finis Welch

Rights and Regulation
Ethical, Political, and Economic Issues
Edited by Tibor R. Machan and M. Bruce Johnson
With a Foreword by Aaron Wildavsky

Water Rights
Scarce Resource Allocation, Bureaucracy, and the Environment
Edited by Terry L. Anderson
With a Foreword by Jack Hirshleifer

Natural Resources
Bureaucratic Myths and Environmental Management
By Richard Stroup and John Baden
With a Foreword by William Niskanen

PACIFIC STUDIES IN PUBLIC POLICY (continued)

Resolving the Housing Crisis
Government Policy, Decontrol, and the Public Interest
Edited with an Introduction by M. Bruce Johnson

The Public School Monopoly
A Critical Analysis of Education and the State in American Society
Edited by Robert B. Everhart
With a Foreword by Clarence J. Karier

FORTHCOMING

The American Family and the State

Stagflation and the Political Business Cycle

Rationing Health Care
Medical Licensing in the United States

Oil and Gas Leasing on the Outer Continental Shelf

Taxation and Capital Markets

Myth and Reality in the Welfare State

Electric Utility Regulation and the Energy Crisis

Immigration in the United States

Crime, Police, and the Courts

Drugs in Society

For further information on the Pacific Institute's program and a catalog of publications, please contact:

PACIFIC INSTITUTE FOR PUBLIC POLICY RESEARCH
177 Post Street
San Francisco, California 94108

CONTENTS

LIST OF FIGURES

LIST OF TABLES

FOREWORD

Despite popular perceptions to the contrary, the painful shortages and price escalations of the 1970s in petroleum and natural gas markets were not the result of market failures. Instead, government restrictions of energy markets were responsible for the gasoline lines, factory shutdowns, and full misallocations of the mid- and late 1970s. In this excellent book, economist David Glasner sets the record straight. He exposes the lack of evidence supporting those who blame the private sector companies, and he carefully details how the energy policies of government caused the American economy to operate less effectively. Moreover, he shows that these faulty policies were not merely the result of honest mistakes and bad judgment, but form a pattern of reaction to normal supply reductions in competitive markets for all types of energy. This timely study of a vast array of economic research provides important lessons for policymakers as we go into another period of supply stringency in the late 1980s and 1990s.

Paul W. MacAvoy
Dean, Graduate School of Management
University of Rochester

INTRODUCTION
The General Effects of Energy Policy

Once upon a time, my dog and I were strolling leisurely through the park when we happened upon a certain gentleman seated on a park bench. While imbibing from a bottle of muscatel, he proceeded to inform me (my dog being quite a skeptic in such matters) that the good earth is indeed flat. Ever a faithful employee of NASA's Jet Propulsion Laboratory, I produced from my pocket a satellite photo of the earth, showing it to be unmistakably round. I discussed the behavior of the tides, of the winds, of objects on the horizon. I explained patiently that these and a multitude of other observable phenomena are consistent only with a round-earth hypothesis. Needless to say, this gentleman could not be swayed from his conviction that, evidence or no evidence, the earth is flat, period. My dog and I proceeded on our way, sadder but wiser, yet more confident in the knowledge that here at least was one individual impervious to the charms of, say, a typical television weatherman.

Let me say now that our ever-vigilant advocates of "energy policy" in its many guises constitute the political equivalent of the Flat Earth Society. Literally no amount of analysis or evidence, however sound or extensive, can disabuse such individuals of the notion that a

Formerly the energy policy senior staff economist for President Reagan's Council of Economic Advisers, Benjamin Zycher is now a senior economist at the Jet Propulsion Laboratory, California Institute of Technology. The views expressed are his own.

"strong energy policy" can reduce prices, inflation, unemployment, poverty, injustice, instability, vulnerability, and—in all probability—baldness. And do not delude yourselves for a moment that such evidence is lacking. On the contrary, as this excellent book makes abundantly clear, the evidence is overwhelming that federal energy policies generally increase true prices, make the poor and consumers worse off, reduce stability in energy markets, increase vulnerability to supply disruptions, generate queues and other forms of chaos for market participants, and in general cause a never-ending sequence of predictable disasters. How is it then, you ask, that energy policy is so permanent a feature of the political landscape? Why is it like a Roach Motel, easy to check into but exceedingly difficult to abandon? Ah, herein lies the secret of energy policy: at its true purpose, specifically the generation of wealth transfers to the politically influential, it is a shining success. Unfortunately, society as a whole is the sad and substantial loser in this game of legalized theft; more about this below.

PURPORTED AND ACTUAL EFFECTS OF PRICE AND ALLOCATION CONTROLS

For those who idled away the last decade asleep (three decades in the case of natural gas) or in Washington, D.C., the following eternal truths are worthy of memorization. These facts of life are analyzed in detail in this book, but a brief summary may be useful at this point.

1. *Price controls do not reduce prices.* If prices are prevented by regulatory fiat from allocating supplies among competing uses and from limiting consumption to amounts available, then *nonprice rationing* must be used. This is unavoidable: *some* mechanism or device must perform the rationing function. In 1973 and 1979, queuing (that is, standing in line) served this role in the gasoline market. The salient point is that the controls reduce only reported prices, which exclude time and other resources squandered in efforts to obtain fuel at artificially low ceiling prices. Under binding gasoline price controls, the relevant price is not the ceiling price alone but is instead the *sum* of the ceiling price and the value of time spent waiting in line. This sum is *unambiguously* higher than the price that would clear the market (without queuing) in the absence of the price

controls. This is because the controls reduce output, thus raising the (marginal) market value of gasoline. Competition for gasoline among consumers induces them to spend an amount of time in line sufficient to exhaust the difference between this higher marginal value and the legal maximum price on the pump. In short, by reducing output, controls raise true prices.

2. *Price controls do not bestow benefits upon consumers or the poor.* The interest of consumers is served not by an artificially low price for a single good but instead by increases in the size of the total consumption basket. By raising true prices, and causing resource waste and allocational inefficiency, controls reduce the total consumption basket, thus making consumers as a class clearly worse off. Moreover, controls lead to increased oil imports due to both reduced domestic production and indirect subsidies, such as entitlements regulations, for users of imported oil. This increased importation of oil weakens the dollar, thus raising the domestic prices of all other imported goods and inducing the U.S. economy to send more domestic goods overseas.

Controls are likely also to hurt the poor for several reasons, the most important of which is the aggregate-wealth effect of controls and the perverse incentives they provide. No one can argue seriously that reduced aggregate wealth makes the poor better off, either directly or in terms of upward mobility across generations and income classes.

More specifically, there are four basic reasons that controls are inconsistent with the interests of the poor. First, price controls on gasoline, for example, constitute a constraint on price rationing of gasoline, that is, on the free market allocation process. In other words, controls constrain the ability of individuals to compete for gasoline in terms of dollars so that competition must take other forms. To the extent that the nonpoor are prevented by such constraints from consuming additional units of gasoline, they will consume additional units of other goods instead, driving up those prices. In short, the oil controls may reduce the ability of the poor to compete for goods in nonoil markets. Nonprice rationing of petroleum thus may induce a demand shift process, making the poor worse off with respect to the entire consumption basket. Furthermore, this shift in relative prices, caused by the artificial constraints imposed by the oil price controls, will induce a series of windfall gains and losses for a wide variety of individuals and groups, both consumers and producers. The net distributional effect of these windfalls cannot be predicted in advance, but some poor individuals would undoubtedly become losers.

Second, constraints on market allocation (that is, price rationing) cannot make the poor better off unless the alternative allocation mechanism itself favors this group. Surely queuing as an allocation device does not obviously favor the poor either in general or with respect to any particular good. This problem *cannot* be avoided; once market allocation is constrained, there *must* be an alternative rationing mechanism for any good the market value of which is greater than zero. In order for the poor to be made better off by queuing, they must have greater willingness than others to spend time in the queue. The poor do not obviously have lower marginal time values than the nonpoor if their schedules are less flexible. Even if the marginal time values of the poor are lower than those of the nonpoor, queuing may still favor the latter group if the relative value of the good to the nonpoor is sufficiently high relative to that of the poor. In other words, if the nonpoor value the good much more highly than the poor, they may be willing to "outwait" the poor just as they would be willing to outspend (or outbid) the poor if prices were used as an allocation mechanism. Nor is allocation by government regulation more likely to achieve the stated "equity" aim: what regulatory process can be expected with confidence to favor the poor? If price and allocation controls are advocated in the name of equity, the alternative allocation mechanism must be specified and shown to be more consistent with the interests of the poor.

Third, price ceilings reduce prices received by producers, and increase true prices paid by consumers. Therefore, ceilings are analytically equivalent to a tax, the incidence of which is likely to fall upon both (some) consumers and the owners of inputs in the production of the good in question. Are the producers of "luxury" goods—or of gasoline—themselves wealthy? Are pension funds and other owners of oil company stocks wealthy? An answer of no is wholly plausible for many of these owners, in which case the group of losers might well include many poor individuals.

Finally, a major vehicle for improvement in the position of the poor—particularly in terms of upward mobility across generations—is economic growth. The interest of the poor lies in a growing economy with expanding opportunities for employment and the acquisition of wealth. Policies that hinder allocational efficiency—that is, price and allocation controls—must have the effect of reducing aggregate wealth due to both direct and disincentive effects. At least part of this burden of reduced wealth would fall on the poor. In addition, reduced aggregate wealth may reduce the amount of transfers supplied to the poor by the nonpoor.

3. *Allocation regulations are not "fair."* Allocation regulations are inevitably based upon some combination of historical use patterns and special provisions designed to placate important voices in Congress. In what relevant sense can it be said that these interests are more "deserving" than others? The bureaucrats are "fair" only from the viewpoint of those able to get cheap oil during disruptions; for others, numerous adjectives exist to describe the inevitable chaos experienced in 1973 and 1979, but "equitable" is not one of them.

4. *Standby price and allocation controls do not provide "preparedness."* Instead, they reduce it. Preparation for contingencies means investment in stockpiles and other assets designed to reduce the adverse consequences of future disruptions. Federal policy must facilitate both the allocation of available supplies to their most productive uses and the production of substitute fuels. Willingness to make such investments is contingent upon an expectation of profit potential commensurate with attendant risks, an environment absent with the expectation of controls. This expectation reduces preparation incentives both for those who expect to see (the value of) their oil confiscated and for those who expect the government to bail them out. The burden of the Strategic Petroleum Reserve now being carried by the taxpayers is a measure of the "benefits of advance preparation" provided by controls.

5. *Simulation of the effects of controls with most macroeconomic models is worthless.* This is so for a number of reasons, the most fundamental of which is that the computer and, often enough, the economist using it do not understand the difference between reported prices and true prices under controls. What the computer reads as a price "reduction" because of controls is actually an increase in the true price; the resulting predicted macroeconomic "benefits" of controls are an utter illusion because controls cannot produce more oil.

These predicted "benefits" include lower inflation, lower unemployment, and enhanced national wealth. Again, these are illusory.

By causing allocational inefficiency, controls must have adverse aggregate impacts. If the market is allowed to clear at the higher postdisruption price, resources are induced to shift among sectors, resulting in short-run resource unemployment and aggregate wealth losses. The relevant question, however, is the expected present value of this wealth loss relative to that borne if resources are artificially induced to remain in their predisruption employments. *The latter (expected) wealth loss is greater.* Resource owners, after all, always have the option of reducing their prices in order to remain in predisruption employments, thus avoiding the short-run unemployment costs of searching for new employment in light of the changed price relationships in the economy. That they do not do so—that they choose to "invest" by searching for new employment—implies that the net expected present value of the resource shift process is positive. In other words, the market expectation is that the postdisruption resource shift process makes society *wealthier.* Most macroeconomic models predict precisely the opposite; any computer programmed to produce such nonsense has, as it were, a fool for a client.

So there you have it: everything you've wanted to know about the silliness of public arguments favoring controls but were too numb to ask. Well, perhaps not *everything*: this book contains a good deal more.

CONTROLS AND "WINDFALL PROFITS" TAXES IN THE CONSTITUTIONAL CONTRACT

A basic purpose of constitutional (that is, *contractual*) government is to constrain the behavior of political majorities, particularly in terms of confiscatory behavior directed at opposition political groups. Members of the majority may implicitly agree to such constraints as a means of acquiring insurance against the confiscatory behavior of others when members of the current majority become members of the minority. At the constitutional level, any given individual, not knowing whether he, on net, will be in the majority or minority, may agree in constitutional contract to refrain from such confiscatory behavior if he finds himself in the majority in order to be free of similar confiscation if he finds himself in the minority. The constitutional contract, therefore, is analogous to a common (unanimous) commitment to refrain from stealing.

Controls are one means by which Congress can confiscate and transfer wealth without use of the explicit tax system. By raising the stakes inherent in the political process, such policies reduce aggregate wealth because winners and losers alike are induced to make increasing investments in attempts to shape political outcomes. Controls, then, are one of a class of policies that tend to weaken the processes and constraints of our constitutional democracy. Similarly, the crude oil "windfall profits" (excise) tax is nothing more than a punitive tax imposed by the political majority upon an unpopular political group, specifically "Big Oil." Again, this breakdown of the constitutional contract leads winners and losers alike to invest additional real resources in efforts to retain or obtain majority status. These resource investments are a pure social waste because they produce only a changed level or direction of politically mandated wealth transfers among groups. They do not increase overall wealth. In addition, the increased resulting taxation is likely to impose additional deadweight inefficiency losses upon the economy.

Breakdown of the constitutional contract thus leads to an increase in the social costs imposed by the tax/transfer negative-sum game emerging from competition within the democratic process. In other words, individuals are induced to shift some investment away from production of increased wealth toward efforts to steal from others or to prevent others from stealing from them. By shunning punitive taxation of politically disfavored groups—such as oil and gas producers—society can reduce incentives for such socially wasteful investments. It is not accidental that such controls and taxes are favored by those whose interests are served by growing governmental power despite the true adverse effects upon consumers as a whole.

AN AGENDA FOR GOVERNMENTAL RETREAT SUGGESTED IN THIS BOOK

What follows are three specific proposals that, if implemented, would go far toward reducing the damage done by federal energy policies.

Natural Gas Wellhead Price Decontrol

Hope must spring eternal indeed in the human breast; accordingly, there are those who still believe that after three decades of governmental meddling in the natural gas market, a new round of such federal intrusion would cure the latest gas market disarray.

The federal government has regulated prices of interstate natural gas (crossing state boundaries) since 1954. Energy markets were stable until the early 1970s when rising oil prices drove up the demand for gas sharply. The price ceilings encouraged consumption and discouraged production, yielding the shortages always observed under any system of price ceilings. Shortages never afflicted the unregulated intrastate market. The cold winter of 1977 together with the price controls produced a massive interstate gas shortage, causing hundreds of thousands of layoffs and numerous other adverse effects. Natural gas reserves fell during the 1970s by 96 trillion cubic feet (tcf)—over one-third of total reserves or almost five years' consumption. The price controls proscribed high prices for domestic producers so that higher prices still were paid to foreign producers instead.

Driven by the 1977 catastrophe, Congress passed the Natural Gas Policy Act (NGPA) in 1978. The NGPA extended controls to intrastate gas and created about two dozen categories of gas—some regulated, some not, some to be decontrolled in 1985, some in 1987, and some never.

The net effect of this hodgepodge is serious inefficiency in the production and consumption of gas. Because of the various price categories of gas, producers are induced to exploit costly gas sources before less expensive ones and generally to produce a mix of gas that does not minimize costs. Gas prices to given consumers are an average of the various categories of gas delivered to those consumers; this means that unequal geographic access to "cheap" gas bestows an artificial advantage in the competition for unregulated supplies upon those with proportionately greater access to cheap gas. Thus, because of the NGPA, gas is both produced and allocated inefficiently.

The NGPA deregulates roughly half of all gas in 1985. Because gas competes with oil, gas deregulated after 1985 will receive prices *above* the market clearing level that would prevail if there were full decontrol, as *average* gas prices are bid up to oil-equivalent (full decontrol) levels. This has happened with the small amount of gas already deregulated under the NGPA. Various institutional factors caused prices to rise substantially over the last two years so that most gas prices, particularly in the industrial sector, are already close to equivalent oil prices. Therefore, existing gas consumers as a group do not benefit from the NGPA controls; instead, various consumers subsidize other consumers, and producers of deregulated gas receive a subsidy from other producers.

Predictions of future gas shortages were common in the late 1970s. These predictions, together with fresh memories of the winter of 1977, induced the gas pipelines to secure future supplies by signing contracts with historically high "take-or-pay" provisions, obligating payment for the gas even if it cannot be resold (until later) to gas consumers. In effect, these contracts constituted insurance for the pipelines and their customers. The take-or-pay obligations explain seemingly irrational behavior: purchase of "expensive" gas when "cheap" gas is available. Moreover, some observers have argued that because prices have been rising during a period of gas surpluses ("excess deliverability"), the gas market must be "noncompetitive." This argument is incorrect. Excess deliverability has existed because the take-or-pay obligations were fixed before the recent gas demand decreases, caused by the past recession and other factors. Prices have been rising because the artificial constraints imposed by the NGPA have eased and because of some institutional factors. "Surpluses" force prices down only if we begin with the market price, not with an artificially low regulated one.

Full decontrol would remove the production and consumption inefficiencies discussed above and would eliminate the subsidies generated by the NGPA. Higher prices for "old" gas would increase economically recoverable reserves of old gas; a conservative estimate of this increase is 5–10 tcf. Prices will be determined by competing oil prices, particularly in the industrial sector, which accounts for about 75 percent of total gas use. Because prices are already close to the oil equivalent level, they would rise little, if at all, upon full decontrol; furthermore, average prices under *existing* policy will equal fully decontrolled prices after 1985 in any case.

Again, the interest of consumers is served not by a low price for any single commodity but instead by maximum output of all goods and services. This requires economic "efficiency," which cannot be achieved under price controls. The goal of "economic efficiency" may seem somewhat sterile, but it is nothing more than a quest for maximum satisfaction of human wants given that not all such wants can be satisfied simultaneously.

Synthetic Fuels Subsidies

Do not be deluded for a moment that only the military prepares for past wars. For we find now efforts by the U.S. Synthetic Fuels Corporation (SFC) to squander yet more billions in taxpayers' money on new synthetic fuels projects. This is, believe it or not, in preparation for the illusory but eternal Energy Crisis, without which whole sections of Washington would resemble a ghost town. The current villain in our capital is "complacency," defined as an unwillingness to subsidize wasteful schemes or otherwise to get the government's snout out of the energy policy trough.

And synthetic fuels subsidies constitute a trough at which only white elephants will feed. The projects envision price guarantees for synthetic fuels of up to $67 per barrel–this in a market with prices at about $29, with only weak prospects for increases. Energy markets would be subject to less distortion and uncertainty, and society as a whole would be wealthier were this program eliminated.

Production of subsidized synthetic fuels will do literally nothing to reduce U.S. energy "vulnerability." This is true for a number of reasons, the most fundamental of which is that the United States is part of the world oil market in which there can be only one price for oil. This means that the effects upon oil prices in the United States of supply disruptions do not depend upon the degree to which we are "self-sufficient." In other words, an oil supply disruption would raise prices in the United States by the same amount whether we import all or none of our oil. Furthermore, even according to plan, the synfuels programs will produce only trivial amounts of oil; a program sufficiently large to affect the world price would be fantastically expensive. Other "preparedness" options, the most important of which is the Strategic Petroleum Reserve, are more effective, far less costly, and available now. In short, the "insurance" purportedly provided by synfuels subsidies is nonexistent, and it imposes upon the taxpayers risks and costs regarded by the market as not worth bearing.

Synthetic-fuels subsidies cannot be justified on the basis of the alleged "riskiness" of the projects. Subsidies cannot reduce risk in any relevant sense but can only shift it onto the taxpayers. This results in a wasteful distribution of risk within the energy sector and the overall economy. Moreover, the market undertakes a wide variety of large and "risky" projects, some energy related, with long lead times, long economic lives, and uncertain futures. Subsidies for synthetic fuels are no more justifiable than would be subsidies for refineries or petrochemical plants.

Sound economic policy should attempt to maximize the value of all output, that is, the satisfaction of human wants. Synfuels subsidies are inconsistent with that goal because large sums are invested in the production of small amounts of very expensive fuel. Why substitute the judgment of bureaucrats for that of the market? Moreover, sound policy should avoid creation of concentrated interest groups with an institutional interest in more expensive energy, in shortages, and in policies yielding such dubious results. The SFC is such an interest group. As part of Washington's Permanent Energy Policy Interest Group (PEPIG), the SFC has an institutional interest in perpetuation of the energy crisis and in more government programs, subsidies, controls, and so on. Furthermore, the SFC acts as a central receiving station for the appeals of private-sector groups with similar interests. Let us not deceive ourselves; there are plenty of them. Public policy should not encourage such supplication but should instead make it more risky and less rewarding.

How long can a system survive in which business firms enjoy profit potential but taxpayers bear downside risk? Inevitably, Congress will demand a quid pro quo, leading to price controls, poorly designed taxes, and other policies yielding poor productivity in the energy sector. In particular, these policies result in reduced private sector willingness to make investments in preparation for supply disruptions. Hence, a supreme irony: synfuels subsidies and similar policies will lead ultimately to *reduced* aggregate preparedness for emergencies.

The Department of Energy

It is imperative that there not exist in our beloved capitol a concentrated interest group, supported by tax dollars, with an institutional stake in a return to regulatory fiat as a policy for dealing with supply disruptions. That is what much of the Department of Energy is, however muted it may be during the Reagan administration. This bureaucracy has an interest in maximization of its work load, authority, and influence. The present market policy serves none of those ends; a price and allocation control apparatus does—beautifully. The Department of Energy serves also as the major central receiving station for the demands and appeals of all the special interests lobbying for regulatory policies favoring them. It is important to remove institutional arrangements that facilitate such political pressure. The Department of Energy is such an institution. In short, abolition of the Department of Energy is important in its own right and also in order to defend market policies.

This fine book deals with these issues and much more in a useful way that is not possible in a short introduction. Read it, then, and marvel at the emptiness of the standard television news "analysis" of energy issues and at the amazing ignorance of most of our public servants. And remember, regardless of the apparent sincerity of their pronouncements, that the earth is indeed round.

Benjamin Zycher
California Institute of Technology

PREFACE

It is not so long since the United States and the rest of the world experienced a series of traumatic events that came to be described as the energy crisis. Oil, which had for decades been cheap and plentiful, suddenly became expensive and rare. And for a few months in late 1973 and early 1974, it seemed as if the moment when supplies of oil would finally be depleted was about to arrive. Indeed, with motorists waiting in line for hours outside gasoline stations hoping to get one more tankful before the pumps ran dry, gasoline appeared to be running out all across the country.

Many explanations were offered for the gasoline shortages and the onset of the energy crisis. Since shortages began to occur after the Arab oil-producing countries announced an embargo on oil shipments to the United States in retaliation for American support to Israel during the Yom Kippur War, it was understandable that many should attribute the shortages to the Arab oil embargo. Others, however, insisted that the entire crisis was simply a conspiracy on the part of the oil companies to enrich themselves at the public's expense. And there were also those who contended that the shortages and the energy crisis were the result of years of wasteful consumption of energy, which had led to the depletion of domestic, and an excessive reliance on unreliable foreign, sources of energy.

But elementary economics offers a very different explanation for the trauma that we were undergoing. According to economic theory, the shortages were caused by the controls that had been imposed on the prices of crude oil and refined products. Although the Arab oil-producing countries were obviously able to raise the price of crude oil by reducing their output, neither they nor the oil companies were in any position to bring about the shortages that obliged motorists to wait in line for hours before they could buy any gasoline. In the absence of controls on prices, market forces would have automatically ensured that whatever supplies were available would have been allocated so as to prevent any shortages from arising. Even the ostensible embargo that the Arabs had announced on oil shipments to the United States was essentially irrelevant as long as Americans were not prevented from paying the prevailing world prices for crude oil and refined products.

But in the emotional reaction to the shortages and the apparent onset of the energy crisis, these very straightforward considerations were almost entirely overlooked. All that seemed to matter was (1) to find someone to blame and, if possible, punish and (2) to initiate bold, costly, and above all, visible programs designed to alleviate the crisis. On neither count did lifting price controls rate very high. To do so would hardly have been a fitting punishment for the oil companies (which, along with the Arabs, were the leading candidates for villainhood); nor would doing so have entailed the increased expenditure of money and resources that were a prerequisite for any program purporting to deal with the energy crisis. Instead, "energy independence" became the rallying cry of a beleaguered administration seeking desperately to divert attention from an unfolding political scandal.

All these reactions to the shortages and the energy crisis seemed to take as axiomatic the proposition that market forces could not be relied on to allocate petroleum and its products efficiently. Indeed, even to suggest that market forces should allocate available supplies of crude oil and refined products was to invite the suspicion that one was either out of touch with reality or an apologist for the oil companies.

There was no shortage of explanations for why the market was incapable of allocating energy resources efficiently. It was argued that the oil companies, together with the international oil cartel, could manipulate petroleum markets for their own selfish purposes at everyone else's expense. It was also argued that since both the demand for and the supply of petroleum were very insensitive to price, the price adjustments required to induce substantial changes in production and consumption would devastate the economy. Furthermore, it was pointed out that since oil reserves were being depleted, conservation was manifestly a necessity. But certainly to rely on market forces to conserve natural resources was like putting the fox in charge of the chicken coop.

Nor, in this atmosphere of extreme skepticism about the efficacy of market forces, was the presumption that market forces failed to allocate resources efficiently restricted to the markets for petroleum and its products. So it was no accident that the onset of the energy crisis was quickly followed by a spate of proposals calling for the government to assume an enlarged role in allocating resources through some form of national economic planning. Since the market system, according to advocates of planning, had either been instrumental in causing the energy crisis or, at best, had been unable to respond in a way that would have mitigated its effects, similar crises were, in the planners' view, inevitable unless national economic planning, which would foresee and take measures to avoid or moderate those crises, were instituted.

Such arguments about the supposed failure of the market system have always seemed completely wrongheaded to me since the most serious consequences of the energy crisis – the gasoline shortages – were obviously caused by price controls and other interventions in the markets for crude oil and refined products and not by market forces, themselves. Indeed, that the consequences of interfering with and suppressing market forces should be used as evidence of their inefficiency and as a pretext for extending the political and bureaucratic controls that had precipitated the energy crisis in the first place over still wider spheres of economic activity has always impressed me with its intense irony.

When I first became interested in the controls on petroleum prices during my days as a graduate student at the University of California, Los Angeles, I was inclined to regard the persistence of controls despite their manifestly harmful effects as evidence of the irrationality of the political process. This view was reinforced when it dawned on me that controls on domestic crude oil prices could not possibly be accomplishing their announced purpose of reducing the prices to consumers of gasoline and other refined products. Basic economic analysis demonstrates that controls on domestic crude oil prices had no tendency whatsoever to reduce the prices of refined products.

The reason is obvious. After March 1974, there were no longer any shortages of refined products, so their prices must have been determined by the forces of supply and demand. Except insofar as it could have affected the supply of products, the domestic price of crude oil was therefore irrelevant to product prices. But price controls on domestic crude oil were certainly not increasing the supply of refined products.

It was from this realization—that controlling the price of crude oil could not have the effect that advocates of controls believed them to be having—that my idea for a study exposing the irrationality of the entire enterprise began to emerge. But as I continued to think about it, I found the notion that the political process operates irrationally increasingly implausible. First, despite the initial interest they generated, proposals for national economic planning were in fact rejected by the political process. Second, by the end of the decade, the political process had begun to relax, and even to abandon, some of the controls on petroleum and natural gas prices.

At a more fundamental level, however, I was becoming conscious of a basic inconsistency between my presumption that the political process operates irrationally and my general approach to economic analysis. In this approach, the assumption that all decisionmakers rationally pursue their self-interest is regarded as a fundamental methodological principle of *universal* application. If this assumption is indeed universally applicable, it must apply to the political process as well as to the marketplace, so the simple assertion that the political process operates irrationally is neither scientifically fruitful nor intellectually satisfying.

As I began to learn more about the manner in which the entitlements program operated in conjunction with the price controls on crude oil, my dissatisfaction with that assertion increased. For it became evident to me that price controls were not necessarily failing to achieve their desired results, though the results may not have been those stated publicly. What I found was that price controls were a means of transferring wealth from oil producers to oil refiners and that the entitlements program was a highly sophisticated means of redistributing that transfer among refiners. This finding drew my attention to some remarkable similarities between the entitlements program and the distribution of import-quota tickets under the mandatory oil import quota program that had been in effect from 1959 to 1973, just one year before the entitlements program was started.

Thus in attempting to understand how controls on energy prices really worked and what effects they really had, I came to realize that the controls were not the result of a colossal misunderstanding of their effects. Rather, controls were the result of purposeful behavior by individuals seeking to advance their own self-interest. Just as market forces are characterized by certain systematic effects that can be perceived in a wide variety of situations, controls must also be characterized by systematic and predictable effects that work to the benefit of specific groups seeking imposition of controls with those very benefits in mind.

This realization, however, poses a fundamental problem I have attempted to deal with but cannot yet claim to have solved entirely to my satisfaction. The problem is the following: If the controls on crude oil and product prices (or any market intervention) had systematic effects redounding to the benefit of specific groups that used the political process to attain those benefits, then in what sense may an economist render a disapproving judgment about those controls? To do so was certainly my intention when I began writing this book, and to be sure there is a lot of evidence that the controls were harmful. But if all people do indeed rationally pursue their self-interest, why would they tolerate the continuation of a policy that was imposing net losses on the community as a whole? If we take account of all the costs of changing policies, including the costs of finding out the effects of various policies and negotiating an agreement to change policies, then clearly members of a community would not tolerate a policy that was imposing net losses on them. If that be the case, then a judgment by an economist that any given policy is inefficient is problematic. The economist is either ignoring relevant constraints on individuals, ignoring relevant costs of changing policies, or implicitly denying that individuals do rationally pursue their own self-interest.

I believe there are ways of escaping this dilemma. One of them may be to assert that the economist has superior information, not possessed by other individuals in the community, about the effects of alternative policies or about changes in constraints (e.g., laws and institutions) subject to which individuals pursue their self-interest. Thus, what economists are saying when they disapprove of a policy is that if everyone else knew what they know, the policy would be altered. But this response is open to the rejoinder that if economists had superior information about the effects of policies, rational decisionmakers would acquire that information and act on it. If the latter do not make use of the information, it must be because it is not worth their while to find out and act on the information.

There are two possible reasons why it would not be worth finding out and acting on information possessed by economists. One is that the information has little value. The other is that it is very costly to communicate this information to noneconomists. If the latter be the reason, there are considerable gains to be derived from more effective communication between economists and noneconomists.

Thus in seeking to communicate knowledge possessed by economists concerning controls on petroleum prices and their effects, I have had two principal objectives. First, to compare the way that markets based on private property rights allocate crude oil, refined products, and natural gas and the way the political process, based on a more ambiguous set of constraints than private markets, has allocated those resources. The information that economists have seems to me to support a judgment that markets allocate resources more efficiently than has the political process. My second, and more difficult, objective has been to explain why—if the market is indeed a more efficient allocator of energy resources than the political process—a rational political process still chose to supersede the market as an allocator of energy resources.

Since I have written this book out of the conviction that communication by economists to noneconomists is neither useless nor impossible, I have sought throughout to make the analysis accessible and interesting to the lay reader to whom this book is primarily addressed. I have used diagrams sparingly, and I employ only one equation—though I admit it is a somewhat nasty one. In two places I have chosen to elaborate slightly on theoretical points. These elaborations are found in appendices to Chapters 2 and 10, respectively. But readers prepared to accept the theory as I present it in those chapters need not trouble themselves with the more extensive development of the theory in the appendices. I have not assumed any prior knowledge of economic theory on the part of the reader. My hope is that anyone seriously interested in the implementation of energy policy during the 1970s and how that policy might be improved during the 1980s will be able to follow my entire discussion without excessive difficulty and even with some interest. The book should also prove useful as a supplementary text in courses on energy economics and energy policy, regulation and public policy, and applications of microeconomics.

Although I have not written this book primarily for fellow economists, the questions it addresses certainly deserve economists' close attention. I cannot claim to have advanced the frontiers of knowledge very far, but I have brought together a wide range of theoretical, historical, and descriptive material from a number of areas that nonspecialists in those fields may find useful and informative.

Part I of this book provides a theoretical framework for understanding the reasons for the superiority of competition by offers of exchange to other forms of competition. Chapter 1 discusses the relationship between the seemingly conflicting concepts of competition and cooperation. I argue that competition by offers of exchange is more conducive to human cooperation than is any other mode of competition. The tendency to promote cooperation is one manifestation of the general tendency of competition by offers of exchange to maximize wealth. In Chapter 2 I consider whether the exhaustibility of petroleum resources vitiates the argument in favor of competition by offers of exchange. One might think so since future generations cannot compete with present ones for the use of exhaustible resources. Consideration of this question leads me into a discussion of the theory of private property rights and their role in the conservation of resources. The upshot of this discussion is that insofar as future exhaustion of resource stocks can be correctly foreseen, private ownership does ensure the appropriate withholding of those stocks for future use. In Chapter 3 further consideration is given to this question without making the assumption of Chapter 2 – that future supplies and demands for resources can be correctly foreseen in the present. I argue that uncertainty about the future does not imply an extremely conservative policy toward use of exhaustible natural resources since future generations can be made worse off by too slow a rate of natural resource use as well as by too rapid a rate. As a general principle, reliance on market forces is likely to generate the most informed decisions about whether natural resources should be used up in the present or conserved for future use.

In Part II I turn to a detailed investigation into the role of restrictions on competition by offers of exchange in bringing about the energy crisis of the seventies. Chapters 4 through 6 deal with the effects of price controls on markets for refined products, especially gasoline. Chapter 4 is primarily concerned with providing historical and analytical background for understanding the unique structural forms of the gasoline market and how it has been evolving since the early part of this century. The background will be helpful in understanding many of the issues raised in the analysis of price controls and other restrictions on competition by offers of exchange in succeeding chapters. Chapter 5 presents an account of the imposition of price controls on the gasoline market, beginning with President Nixon's announcement of a freeze on all wages and prices in August 1971. The chapter goes on to explain how the gasoline shortages that followed the announcement of the Arab oil embargo against the United States in the fall of 1973 were, in fact, caused by the controls and not by the embargo, itself. Chapter 6 presents a theoretical model of the operation of controls on gasoline prices that explains how controls allowed gasoline prices eventually to rise to market-clearing levels so that, for the most part, shortages of gasoline disappeared after March 1974. The chapter concludes with an explanation of why shortages reappeared again in the spring and summer of 1979.

Chapters 7 and 8 provide a similar historical and theoretical account of controls on crude oil prices. In Chapter 7 I discuss the evolution of the domestic market for crude oil and describe both the development of demand prorationing that for many years was carried out by individual states, most notably Texas, and of the oil import quota system. This is followed by an account of the breakdown of the import quota system in the early seventies and its almost immediate replacement by a system of price controls on domestic crude oil and an entitlements program that preserved many of the essential features of the import quota system. These similarities suggest that despite obvious differences between the two programs, many of the political forces that influenced the quota system were equally effective in shaping the price controls on domestic crude oil. In Chapter 8 I explain why the widely held presumption that price controls on crude oil held down the prices to consumers of gasoline and other refined products could not have been true. I contrast the naive cost-of-production theory on which this presumption is based with the marginalist theory of standard economic analysis, which implies that price controls on domestic crude oil prices could not, by themselves, have reduced the prices of refined products. The entitlements program introduces a certain ambiguity into the analysis, though, since it had a tendency to reduce product prices at any given level of world crude oil prices. However, since the world price of crude oil was probably increased significantly by the entitlements program, the overall effect of the price controls together with the entitlements program was probably to increase the consumer price of gasoline and other refined products. Empirical evidence at the end of Chapter 8 corroborates this conclusion.

Chapters 9 and 10 deal with two remaining issues for future economic policy: the windfall profits tax and controls over the price of natural gas. In Chapter 9 I use the results of the previous chapter to challenge the notion that windfall profits accrued to the oil industry as a whole because of decontrol. Moreover, by discouraging domestic oil production and reducing its responsiveness to changes in the world price of oil, the windfall profits tax tends to raise the world price of crude oil. After discussing the use of revenues from the windfall profits tax to finance synthetic fuel production, I conclude the chapter with a discussion of the concept of a windfall profit and the moral argument for taxing away such profits. In Chapter 10 I take up the question of controls on natural gas prices, reviewing the history of the natural gas market and the evolution of controls over the price of natural gas at the wellhead. After documenting the responsibility of controls for natural gas shortages during the seventies, I go on to show how the current complex system of controls discourages production for low-cost old gas and subsidizes the production of gas from new high-cost sources without holding down the prices to consumers. Indeed, an argument similar to the one in Chapter 8 illustrates how controls on natural gas prices are helping to support the world price of oil and how termination of controls could lead to a further general reduction in energy prices similar to that which occurred after the decontrol of domestic crude oil prices.

Part III consists of a single chapter containing some concluding thoughts about politics, markets, and energy. I review the leading principles developed in Parts I and II and address several arguments that are still made rather frequently in favor of government controls on energy prices and allocation of supplies.

ACKNOWLEDGMENTS

In writing this book, I have received assistance from many individuals whose kindness I should like to acknowledge. Charles Breeden, Douglas Den Uyl, Rhonda Hageman, Scott Harvey, Thomas Hogarty, Calvin Roush, Susan Shapiro, Julian Simon, and Peter Toumanoff all read drafts of one or more chapters and made numerous helpful comments and suggestions. I am particularly indebted to Harvey and Roush for their analysis of price controls on refined products, which I only inadequately understood before reading their important study of this complex topic. James Johnston was a constant source of encouragement and of information about markets for crude oil, refined products, and natural gas. I also benefited from discussions with Rodney Smith, Alan Reynolds, and my colleagues in the economics department at Marquette University. Johnston, Charles Baird, Gary Libecap, and Tim Ozenne all read the entire first draft of the book, and I very much benefited from their incisive comments and criticisms. Such errors and shortcomings as remain are, of course, no one's responsibility but my own.

I also want very gratefully to acknowledge the financial support I received from the William L. Law Foundation and the Earhart Foundation while I was working on this book. Thanks go as well to the College of Business Administration at Marquette University for summer research funds. And I owe a particular debt of gratitude to Dean Thomas Bausch of the College of Business Administration and to my department chairman, Brian Brush, for their understanding in permitting me to take a leave of absence so that I might devote full time to writing.

Intellectual debts are the most difficult to acknowledge because it is so hard to identify the sources of and inspirations for one's ideas. There are, I am sure, more such debts left unmentioned, but of the many that I owe, I want at least to acknowledge those to my professors at UCLA, especially Armen Alchian, Harold Demsetz, Jack Hirshleifer, Benjamin Klein, Axel Leijonhufvud, and Earl Thompson, who taught me most of what I know about economics and from whom I hope to continue learning for a long time to come. And those familiar with the work of F. A. Hayek will readily see the enormous influence he has had on my thinking.

For their many personal kindnesses that in one way or another helped me finish this book, I must also thank Nyle Kardatzke, Ruth Schmelzer, Menachem Schmelzer, and my parents. For providing me with research assistance, I want to thank Chet Ceille, Daniel Persons, and Robert Hoffman. Jay Stewart helped greatly with her editorial assistance; Naomi Schmelzer, Carl Moseson, and Angela Koutoulakis were diligent proofreaders. Ronni Bellet, Brenda Gandelman, Audrey Lemanske, Kathy Lentsch, Mary Menasian, Alannah Orrison, Charles Steele, and Mary Voell all helped in typing either the final manuscript or its many earlier incarnations. The fact that I plan to write my next book on a word processor should in no way be taken as a reflection on their performance.

PART I

THE THEORETICAL FRAMEWORK

Chapter 1

SCARCITY, COMPETITION, AND COOPERATION

There exists in every human breast an inevitable state of tension between the aggressive and acquisitive instincts and the instincts of benevolence and self-sacrifice. It is for the preacher, lay or clerical, to inculcate the ultimate duty of subordinating the former to the latter. It is the humbler, and often the invidious, role of the economist to help, so far as he can, in reducing the preacher's task to manageable dimensions. It is his function to emit a warning bark if he sees courses of action being advocated or pursued which will increase unnecessarily the inevitable tension between self-interest and public duty; and to wag his tail in approval of courses of action which will tend to keep the tension low and tolerable.

D. H. Robertson*

SCARCITY AND COMPETITION

The fundamental postulate of all economic—and for that matter all biological and evolutionary—reasoning is that resources are scarce.

*The epigraph for this chapter is taken from Robertson, *Economic Commentaries*, pp. 148–49. It is hard to resist quoting the final paragraph of Robertson's essay:

> Let me draw on (one) of my favorite sources of wisdom to summarize what I have been trying to say, and to suggest an answer to the question posed in my title—What does the economist economize? "Tis love, tis love," said the Duchess, "that makes the world go round." "Somebody said," whispered Alice, "that it's done by everybody minding their own business." "Ah well," replied the Duchess, "it means much the same thing." Not

The very notion of choice between alternatives, which is the essence of economic theory, implies a sacrifice of something valued. In a world in which resources were not scarce, *all* wants could be satisfied, so there would be no alternatives to compare, no choices to weigh, no sacrifices to incur, and no costs to bear. Economic behavior in such a world would, in short, be inconceivable.

The same is true of evolutionary theory. If the capacity of the world to support life were not somehow limited, that is, if resources were not scarce, the central ideas of evolutionary theory—adaptation to an environment and natural selection—would be meaningless.

That is not the kind of world we live in. We all have to choose between various good things we should like to have more of than we do. Since each of us has a unique identity with distinct tastes, concerns, and values, our choices inevitably place us in competition with each other. Even selflessness and altruism are not enough to eliminate competition for the control and use of scarce resources.

I stress this point because there is a tendency to assume that it is only our selfish instincts that drive us to compete. My contention is that once there is more than one individual who can influence the way resources are used, competition is inherent in the very nature of scarcity. Even if we were all altruists, unless we could somehow reach a perfect consensus on what goals to pursue and how to pursue them, we would still be competing. Any means by which we determine how the resources at our disposal are going to be used necessarily implies some competitive process. Thus although competition may be manifested in a number of subtle ways, all societies are competitive.

To illustrate: Given our available resources, we may want to spend money on a number of worthy causes. For example, we may want to educate our children so that they will be able to read books, enjoy great literature, study history, philosophy, and the sciences. We may want to build hospitals to care for the sick and injured. And we may also want to have wide open spaces so that we can enjoy our natural environment. These are all worthwhile and laudable objectives. But even though we all agree they are worthy goals, some of us will want to spend more on education than others, who would rather increase

---

perhaps quite *so* nearly the same thing as Alice's contemporaries thought. But if we economists mind our own business, and do that business well, we can, I believe, contribute mightily to the economizing, that is to the full but thrifty utilization, of that scarce resource love—which *we* know just as well as anybody else, to be the most precious thing in the world.

spending on hospitals or parks. We will, in other words, be in competition with each other. The competition is not due to selfishness, bad faith, or insensitivity. It is merely the unavoidable consequence of differences of opinion or values in a world of scarcity.

Now it may be true that since people are not entirely altruistic and have selfish motives competition is more intense than if everyone were entirely altruistic. But even that is not obvious since people sometimes are most determined to have things their own way when fighting selflessly for what they regard as a just and worthy cause.

If competition really is inevitable, doesn't that make cooperation impossible? Yet people obviously cooperate as well as compete. How, then, is it possible for cooperation to exist in a world of scarcity and competition? And doesn't the existence of cooperation refute my contention that competition is both inevitable and pervasive?

COMPETITION AND COOPERATION

We humans are almost completely dependent on each other for our survival and for satisfying our wants. None of us is, or can be, truly self-sufficient. Indeed, our humanity consists as much in our need for the cooperation of other human beings as in any other attribute. Our need to cooperate with each other, even though we cannot avoid competing, is one of the ironies of the human condition. All social institutions are a reflection of this irony and, in some way, are all meant to help us resolve the tension between the inevitability of competition and our need for cooperation.

The compatibility of cooperation with competition depends crucially on the kinds of competition that are tolerated within a society. How people compete depends on a wide range of factors, such as the traditions, customs, laws, and other institutions that have evolved within a particular society. There may be other influences, as well. But one can at least say that changes in the laws and institutions of a society affect its development precisely by altering the manner in which people go about competing with each other for the means with which to pursue their ends.

The notion of competition seems antithetical to that of cooperation because we tend to think of cooperation in only one of two possible senses. In the first and more familiar sense, cooperation refers to collaboration between two or more individuals in the pursuit of a common goal. A more descriptive term for this sort of cooperation is "teamwork." Since people compete with each other precisely because they don't agree about what goals to pursue or how to pursue them, cooperation (in this sense) is indeed the antithesis of competition. Given the variety of human goals and aspirations, such cooperation could not be very extensive.

Economists recognize another sense in which people can be said to cooperate with each other. None of us consumes only what he or she produces, and most of us consume only what others have produced. That is to say, we specialize in production. The more we specialize, the more dependent on others we become, and hence the more we must cooperate. Adam Smith was probably the first to call attention to the vital role of specialization, which he called the division of labor, in raising the standard of living for large numbers of people:

> It is the great multiplication of the production of all the different arts, in consequence of the division of labour, which occasions, in a well-governed society that universal opulence which extends itself to the lowest ranks of the people. . . .
>
> Observe the accommodation of the most common artificer or day labourer in a civilized and thriving country, and you will perceive that the number of people of whose industry a part, though but a small part, has been employed in procuring him this accommodation exceeds all computation [I]f we examine . . . all these things, and consider what variety of labour is employed about each of them, we shall be sensible that without the assistance and cooperation of many thousands, the very meanest person in a civilized country could not be provided even according to, what we very falsely imagine, the easy and simple manner in which he is commonly accommodated.[1]

What is most remarkable about the sort of cooperation Smith was talking about is that it doesn't seem to require any agreement between producers and consumers on common ends or goals—which is why we don't even think of it as cooperation. If it were only possible to cooperate when there was agreement about the ends we were seeking to accomplish, then specialization and the division of labor would probably not be possible. Producers would have to know the specific purposes for which their products were going to be used. In fact, producers generally neither know nor care about these purposes except insofar as such knowledge enables them to satisfy consumer demands more effectively.

1. Smith, *Wealth of Nations*, pp. 11-12.

If the division of labor does not originate from purposeful cooperation and teamwork directed toward common goals, then what does it originate from? To this question Adam Smith offered a famous response:

> This division of labour, from which so many advantages are derived, is not originally the effect of any human wisdom, which foresees and intends that general opulence to which it gives occasion. It is the necessary, though very slow and gradual, consequence of a certain propensity in human nature which has in view no such extensive utility: the propensity to truck, barter, and exchange one thing for another.[2]

The propensity to truck, barter, and exchange—the willingness to give up one thing of value for another—manifests itself only because scarcity forces us to make choices between a variety of desirable objectives. In making these choices, we always strive to minimize the sacrifice we have to bear to achieve a more desired objective. Thus in pursuing our objectives through trade, we look for the most favorable possible terms on which to trade. But since we all want to trade at the best possible terms for ourselves, we also have to compete for trading partners. We do this by making offers of exchange at terms no worse than other potential traders are offering. I am going to call this type of competition "competition by offers of exchange (COE)."[3]

Smith showed that the cooperation on which we all depend for the maintenance and improvement of our standard of living does not stem from any agreement about the ends we want to pursue but is rather the unintended byproduct of a particular form of competitive behavior, namely, COE. Just as no agreement on goals is necessary for this cooperation to arise, neither is any benevolence or altruism. Cooperation of this kind can emerge simply from the pursuit of individual self-interest.

> Man has almost constant occasion for the help of his brethren and it is vain for him to expect it from their benevolence only. He will be more likely to prevail if he can interest their self-love in his favour, and show them that it is for their own advantage to do for him what he requires of them. Whoever

2. Smith, *Wealth of Nations*, p. 13.

3. I have borrowed this expression from Alchian and Allen, *University Economics*, pp. 11–13. On the pervasiveness of competition in all social settings see Alchian, *Economic Forces*, p. 13.

offers to another a bargain of any kind, proposes to do this. Give me that which I want, and you shall have this which you want is the meaning of every such offer: and it is in this manner that we obtain from one another the far greater part of those good offices which we stand in need of. It is not from the benevolence of the butcher, the brewer, or the baker, that we expect our dinner, but from their regard to their own interests. We address ourselves not to their humanity, but to their self-love, and never talk to them of our necessities but of their advantages.[4]

By making cooperation possible without agreement about objectives or altruism, COE multiplies the opportunities for cooperation enormously. Without COE, my ability to realize my goals is sharply limited. If, as is likely, I cannot achieve them myself, I must enlist the aid of those who either share my goals or wish me well. But the resources and knowledge I can command in this way may be quite unsuited to achieving my objectives. My knowledge of economics, history, and mathematics and my friends' knowledge of music, law, and philosophy will not put food on my table or clothes in my closet. To accomplish this, I need knowledge and resources neither I nor my friends have. COE puts the requisite knowledge and resources at my disposal even though those with the knowledge and resources do not share my goals or wish me well. But to gain access to their knowledge and resources, I have to put what knowledge and resources I have at the disposal of still others whose objectives I may not share or approve of either.

It is in this way that COE allows us all to convert "the things we have and the things we can into the things we want and the things we would."[5] COE thus leads both to a division of knowledge and to a division of labor. Though each of us alone knows how to do very little, by specializing together, we know how to do a very great deal indeed.[6] And the only limit to my ability to use the knowledge and resources of anyone else is my ability to enter into an exchange relationship with that person. Or as Adam Smith put it: "The division of labor is limited by the extent of the market."[7]

4. Smith, *Wealth of Nations*, p. 13.

5. Wicksteed, *Political Economy*, pp. 165–66. Many of the ideas presented in this chapter were developed by Wicksteed in his Chapter V, "Business and the Economic Nexus."

6. See Hayek, "Use of Knowledge in Society," reprinted in Hayek, *Individualism*, pp. 77–91, for a classic exposition of the role of the price system generated by COE in communicating information from widely dispersed sources to those who can make use of it most effectively.

7. Smith, *Wealth of Nations*, p. 17.

WHY DO WE COMPETE THE WAY WE DO?

COE is of course only one of the many possible ways in which people can and do compete. One of the oldest and most common methods has been violence. But because violence is so destructive of valuable human and nonhuman resources, all societies have imposed various internal restrictions on the resort to its use. These restrictions are embodied in the various customs and rules of law and morality that have evolved in all societies.

In fact just about every rule of law or morality can be thought of as some sort of limitation on how people may compete with one another. But when some modes of competition, such as violence, are ruled out or circumscribed, other modes of competition must emerge in their place. And if the new forms of competition are also restricted, still other forms of competition will take their place in turn.

Consider a Hobbesian state of nature in which there are no rules of conduct at all. Competition among individuals would take a variety of forms such as violence, stealth, dissembling, guile, perhaps personal attractions, and so forth. Life in such a state would be, as Hobbes put it, "solitary, nasty, brutish, and short."

As rules governing conduct and limiting the resort to various forms of competition began to evolve, new forms of competition would have to emerge in place of those that had been restricted. If people could no longer fight to gain control over a plot of land, there would have to be some other means of determining how the land will be used. Suppose that the new rules delimit private property rights over the land that allow the owner of those rights to determine the use of the land and to exclude all others from any use of the land without his permission. If these rights are also freely transferable, then competition for the land can be carried out through offers of exchange, and the land could be owned by the person willing to give up the most in exchange for it. But if the land is not freely transferable, then—depending on the means permitted for obtaining control over the land—a variety of modes of competition for using the land may emerge.

To get some idea of what the alternative modes of competition might be like, let us consider a slightly different example, with which most of us have some personal experience. In our society the rules that define and enforce property rights and facilitate their voluntary transfer allow COE to operate over a pretty wide range of goods and services. When COE operates without restriction, a market price emerges so that everyone who wants to buy or sell at that price can buy or sell as much as he wants to. This price is called a market-clearing price.

Gasoline is a good that is usually allocated in this way. Sellers compete by offering gasoline and ancillary services, while buyers compete by offering to pay for these combinations of gasoline and services. But during the seventies this competition was restricted because price controls prohibited sellers of gasoline from accepting offers that exceeded legally prescribed ceiling prices.

To find the effects of these restrictions on COE, two cases must be considered. The first was when the market conditions were such that the market-clearing price was no more than the legally imposed ceiling price. Buyers then could buy and sellers could sell as much as they wanted to at the prevailing price. Despite the controls, this was the normal state of affairs during the seventies, so COE was not really restricted (though it may have been distorted in subtle ways).[8] It was largely unnecessary, therefore, to resort to other forms of competition to determine the allocation of gasoline during such periods.

The second case occurred when the market-clearing price exceeded the ceiling price. The controls prevented buyers who couldn't obtain as much as they wanted to at the ceiling price from competing for the available gasoline by offering to pay more than the ceiling price. If they had been allowed to do so, those willing to pay the most for gasoline would have obtained gasoline that was consumed instead by those less willing to pay for it. The upshot of such competition, let me repeat, is always the establishment of some market-clearing price at which people can buy as much as they want to purchase at that price. This is just basic economics. But what I want to show here is how other forms of competition were resorted to as COE was restricted.

What were the alternative modes of competition? One might be a reversion to violence, but for COE to have been possible at all, violence must already have been circumscribed, if not entirely sup-

8. In Chapter 6 I explain that even when the ceiling price was equal to or above the market clearing price so that final consumers did not have to resort to modes of competition other than COE, price controls were in fact inducing refiners and marketers to alter their behavior in response to special incentives created by the controls.

pressed, by laws protecting person and property. None of these laws, naturally, was repealed. Since one of the alternatives to violence as a means of competing for gasoline had been eliminated by price controls, however, one might conjecture that at least some people who had not done so before began to resort to violence. My impression, based on casual observation, is that the use of violence did increase during such periods.[9]

Nevertheless, other modes of competition were certainly more prominent than violence in taking the place of COE. Consider the situation from the point of view of someone selling gasoline. Since there was an excess demand for gasoline at the ceiling price, the seller would offer less attractive terms to buyers than he had previously. The law prevented the seller from asking for a higher money price, so to achieve the same objective less efficiently the seller offered fewer ancillary services along with the gasoline, no longer checking under the hood or cleaning the windshield. Nor would he exert himself for the convenience of his customers by staying open as many hours as he was accustomed to when COE was not restricted.

While sellers had less incentive to compete for patronage, customers—effectively prevented from increasing their offers of exchange—sought other ways to compete for the favor of sellers. The most noticeable way in which they did so was to line up in front of gasoline stations before they had opened for business in order to be able to buy gasoline when sellers found it convenient to sell. But customers competed not only in their willingness to wait in line; they also competed in their capacity to be appealing customers to sellers. When COE is unrestricted, it is rare that one's personal characteristics will have any impact on one's ability to buy or sell most goods. Anyone willing to pay or accept the going market price will find a trading partner. But when COE is restricted, the personal characteristics of buyers and sellers increase in importance. Since buyers of gasoline could not offer to increase the price they would pay for gasoline, they attempted to be better customers. Regular customers are better customers than transient customers, so regular customers found it easier to obtain gasoline from their regular suppliers than did transient customers. Attractive members of the opposite sex are better, or at least more pleasing, customers, so they were favored by sellers of gasoline. Other individuals with desirable personal charac-

9. See for example *Time*, 2 July 1979, pp. 19, 22–27.

teristics, for example, appearance, friendliness, blood or family ties, personal status, could also compete effectively in this manner.

This example shows that restricting offers of exchange has two main effects. First, it reduces the willingness of potential trading partners to make reciprocal offers of exchange. If a ceiling price is placed on gasoline, gas station owners decrease the number of hours they stay open and the amount of service they provide. Second, it induces those competing for the available supply of a good to engage in forms of competition other than COE in order to obtain some of that good.

The alternative forms of competition, broadly speaking, are of two types. One type involves competitive sacrifices—an expenditure of resources—on the part of those competing. Thus motorists had to compete for gasoline by queuing up outside gasoline stations. A competitive sacrifice, in some (but not all) ways analogous to a competitive price, was extracted from everyone in order to obtain any gasoline. The other type of competition involves using personal characteristics to appeal to whomever has the authority to allocate the available supply. Those with the right characteristics can avoid having to make the competitive sacrifice extracted from those without the right characteristics. This sort of competition takes place at a number of levels. At one level male service station operators favor attractive female customers, while at another the government allocates additional supplies to politically powerful groups such as farmers. But insofar as these characteristics may be acquired through some expenditure of resources or alteration of behavior and insofar as the choice of characteristics to be favored can be influenced by the expenditure of resources (e.g., by lobbying Congress), the second category of competition begins to merge into the first category.

HOW COMPETITION BY OFFERS OF EXCHANGE MAXIMIZES WEALTH

My argument so far can be summarized by five simple propositions. (1) Resources are scarce. (2) Scarcity implies competition for the use of resources. (3) Cooperation enables us to achieve our ends more readily than we could acting alone. (4) COE allows us to cooperate even as we compete. (5) Restricting COE necessarily leads to the emergence of other modes of competition in its place. But this argument does not take us quite as far as I want to go because I also want to argue that COE has a tendency, not shared by any other mode of competition, to maximize wealth. In later chapters I will demonstrate this proposition in detail through an analysis of the markets for crude oil, refined products, and natural gas and of what happens in these markets when prices are controlled. But I want to explain in a general way why COE tends to maximize wealth.

I am of course prepared to concede that wealth maximization is not necessarily the ultimate criterion for judging the success of a society or any of its institutions. There may be ethical, political, philosophical, and aesthetic values that we should also like to consider before rendering such a judgment. But if, everything else equal, it is better to maximize wealth than not to do so—and I think it is—then the proposition that COE tends to maximize wealth and that other modes of competition do not is by no means trivial.

In a world of scarcity, all economic values are relative. They reflect trade-offs people are willing to make or sacrifices they are prepared to incur. Thus if people compete for goods by offering to pay for them, every good or resource is likely to wind up in the possession of whomever is willing to pay the most for it. In other words, under COE, resources go to those who place the highest value on them. Since wealth is simply the total value of all human and nonhuman resources, wealth is necessarily maximized when resources are put to their most valuable uses.

Valuations, however, do not remain constant. They change as knowledge of new resources, new technologies, and new uses changes. As a consequence, the entire range of valuations is never fully known, and wealth is never literally maximized. But there is always an incentive to find a more highly valued use for a resource than its current one. If, for example, I can discover a resource that is undervalued, I can acquire at least part of the difference between its current value and its potential value by effecting the transfer. This is what the owner of the New York Yankees, Jake Rupert, did in 1919 when he bought a star pitcher from the Boston Red Sox for $100,000 and made him a full-time outfielder for the New York Yankees. The pitcher's name was Babe Ruth. The difference between a resource's current and its potential value represents a profit opportunity that, under COE, can be exploited by buying low and selling high. All profits under COE arise from such transactions, and the literal maximization of wealth would occur when and only when no further profit opportunities were available.

WHY OTHER MODES OF COMPETITION FAIL TO MAXIMIZE WEALTH

The tendency of COE to maximize wealth as well as the failure of other modes of competition to do so may be attributed to two characteristics of COE not shared by other modes of competition. One of these characteristics is that COE requires that those who compete for a resource reveal and communicate to others the value they place on that resource. This comes about because no one can acquire a resource under COE except by offering something in exchange, and those who already possess the resource must consider how much they are willing to forego in order to retain possession. Every offer is a measure of the value placed on the resource by the one making the offer, while the person receiving the offer reveals, by accepting or rejecting it, whether she values the resource more or less than that offer.

Other forms of competition, such as competition by waiting in line, can also reveal how much those competing for the resource value it. But the revelation is both less explicit and less easily communicated when valuations are measured by waiting time than when they are measured by a commonly understood offer of exchange or by a market price.

The second characteristic of COE is that the competitive sacrifice extracted from anyone who acquires a good or resource (in other words the price) is incurred only by that individual. Since the price one person pays is received by someone else, there is no net sacrifice to the community—no resources are expended—in the process of COE. But every other mode of competition that extracts some competitive sacrifice—and consequently has some tendency to allocate goods and resources to those who value them the most—does so without transferring that sacrifice (e.g., time spent waiting in line), to anyone else. COE minimizes the amount of resources used up in the process of competing for resources.

It is true that some competitive processes do not necessarily call for any competitive sacrifice. This happens when competition is based entirely on characteristics independent of individual behavior. But such forms of competition are rare. Moreover, if no competitive sacrifice by individuals is called for, there is no tendency whatsoever for individual valuations to be revealed and, thus, no tendency for resources to be allocated to higher rather than lower valued uses.

CONCLUSION

In this chapter I have attempted to show how COE engenders cooperation among individuals seeking to advance their own interests. This cooperation takes the form of a division of labor that has evolved into a vast and almost unfathomably complex network of mutually contingent activities through which almost all the goods and services we consume are produced and distributed.

For this network to operate at all efficiently, there obviously must be some means for bringing the activities that constitute the network into mutual adjustment with respect to each other. If General Motors is to produce cars, other firms have to produce the appropriate numbers of spark plugs, tires, hubcaps, rear-view mirrors, and so on. Without some coordination between these more-or-less independent activities, the entire network is liable to break down.

Since the network is far too complex for any single mind or group of minds to comprehend in detail, the individual activities cannot possibly be determined or controlled by a central authority. The very complexity of the network requires that it must be left, in large measure, to operate automatically. But if there is no central authority to oversee the separate activities constituting the network, and if those activities are undertaken entirely (or even primarily) with a view to advancing private interests, what ensures that those activities will in fact be mutually consistent and that the network will not break down?

The answer of course is that by and large COE ensures their compatibility. The competition of buyers and sellers leads to the emergence of market-clearing prices for all commodities. At these prices buyers can buy and sellers can sell as much as they want to buy or sell. If someone is unable to buy or sell as much as she wishes at the prevailing price, her offer to pay more or to accept less leads to a change in the price that restores the balance between supply and demand. Not only does COE ensure the coherence of this network of activities, but it also tends to maximize the total value of resources and the total value of the output generated within the network.

Although this result is automatic in a sense, it is certainly not inevitable. For COE will not operate—or at least will not operate efficiently—unless other inconsistent modes of competition are suppressed. The suppression of conflicting modes of competition requires that there be an appropriate legal framework that defines and enforces property rights to all scarce resources and facilitates their voluntary exchange. But a legal framework is never established once and for all. It too is an evolving system. Just as competitive forces influence both the allocation of resources within an economic system and the evolution of that system, competitive forces also influence the nature of the legal framework. In other words, since individuals are not indifferent between alternative specifications of their rights and obligations under the legal system, they compete to alter the legal framework to advance their interests. Democracy is a means through which this competition to affect the legal framework can be carried out without resorting to violence.

The point here is that COE never operates in a vacuum. It always operates within a wider context that delimits its operating range, and the legal context within which it operates is in turn shaped by competitive forces.

Let me give an illustration of this relationship between COE and competition to affect the legal framework, which I shall call political competition (PC). Under COE, every change of circumstances—such as a change in resource availability, a change in technology, or a change in the preferences of the public—gives rise to a series of price changes in various markets. Such price changes benefit some individuals and harm others. Those who suffer losses owing to unfavorable price changes often seek to substitute a mechanism other than COE for determining the distribution of wealth. For example, domestic producers faced with increasing foreign competition do not philosophically accept the judgment of the marketplace, as Adam Smith would have urged them to. They appeal to their legislative authorities to modify the property rights permitting domestic consumers to buy as much as they want of the competing foreign product. Various alterations of property rights would serve their purposes, but perhaps the simplest is a tariff on each competing imported product.

Now it is easy for economic theory to confirm what the argument of this chapter has already suggested: namely, that the tariff reduces (even without considering the effect on foreign producers) the total wealth of the community. That is, the aggregate loss the tariff imposes on consumers exceeds the gain it confers on producers.

In a sense therefore COE gives us the "right" result, and PC gives us the "wrong" result. But why does PC yield the wrong result? Why should the community consent to a tariff if the tariff will diminish its wealth? It is fairly clear that producers are better organized than consumers and can, therefore, compete more effectively in the political process than consumers can. Thus the information that the aggregate loss to consumers from the tariff exceeds the gain to producers is not really communicated to or taken into account by decision-makers under PC.[10]

What I wish to emphasize here is that PC is an alternative for groups dissatisfied with their share of the community's wealth under COE. The greater the amount at stake, the greater the incentive to resort to PC. The more cohesive, homogeneous, and better organized any interest group may be, the lower its costs of engaging in PC.

But in contrast to COE, which tends to maximize the wealth of the community, PC tends to reduce it. This is so first because the competitive sacrifices incurred when engaging in PC are not (except in cases of straightforward bribery) transferred to others as are sacrifices under COE. Second, insofar as PC leads to a suppresison of COE, it prevents resources from being allocated to their most valuable uses while giving rise to still other, less efficient, modes of competition to allocate resources no longer allocated by means of COE.

This book is largely about the energy crisis that the United States endured during the seventies. The essence of what happened then is more or less implicit in the argument of this chapter. Faced with a rapidly rising world price of oil, those who under COE would have suffered losses owing to rising oil prices resorted to PC to avoid or reduce those losses. Some, but not all, did manage to protect them-

10. One could further inquire as follows: If producers can compete more effectively than consumers in the political process, why don't they just effect a straightforward wealth transfer from the rest of the community to themselves without interfering with the price mechanism instead of reducing the total wealth of the community through imposition of the tariff? One would, I think, have to explain the preference for the tariff over the straightforward transfer as the consequence of a greater degree of resistance by the rest of the community to an explicit transfer than to an indirect one via the tariff. A whole range of arguments can be advanced in favor of a tariff to distract attention from the wealth transfer that it entails, which would not be available to justify an explicit transfer. Moreover, a tariff also allows the problem of how to divide the proceeds of the transfer among the recipients – in particular, how to divide it between capital and labor – to be avoided. It would be more difficult to avoid this issue in the event of a straightforward transfer.

selves in this way. But PC did more than prevent market forces from redistributing wealth. The outcome of PC in this instance was a system of price controls that distorted and at times virtually suppressed COE. Distortions of COE led to a general reduction of wealth, since resources could not be put to their most valuable uses and even led to widespread shortages that caused breakdowns in the network of interrelated activities that constitutes our economic system. It was such breakdowns that came to be identified with the energy crisis. Thus the energy crisis had hardly anything to do with any of the factors – the Arab oil embargo, the fall of the Shah, the overdependence on foreign oil, or a conspiracy by the oil companies – usually held responsible for it. The breakdowns and the energy crisis were entirely the result of the suppression of COE.

In Part II of this book, I shall describe in some detail how PC caused these breakdowns by leading to a suppression of COE. Even when no breakdowns were noticeable, distortions in COE had various harmful effects that I shall also explain. But in the remainder of Part I, I want to address the widely held view that there is something special about energy resources – specifically, that those most widely used are exhaustible – which makes it inappropriate to subject them to COE.

Chapter 2

ARE WE RUNNING OUT OF OIL?

The desire of acquisition is always a passion of long views. Confine a man to momentary possession, and you at once cut off that laudable avarice, which every wise State has cherished as one of the first principles of its greatness.

Edmund Burke*

INTRODUCTION

The general argument that COE is efficient and tends to maximize wealth seems compelling to me. Nevertheless, general arguments often admit of exceptions, and by the time allowances are made for all the exceptions, very little may be left of the general argument. One might suggest that this is true of the efficiency argument for COE. I cannot of course discuss in this chapter or this book all the possible exceptions suggested to that argument. There is, however, one alleged exception that I must address since, if valid, it would undermine the case for having COE allocate oil, natural gas, and other energy resources.

The alleged exception is that even if COE does allocate resources efficiently among *current* users of those resources, that doesn't mean

*The epigraph to this chapter is from Edmund Burke, "Tract on Popery Laws," in *Works*, p. 352.

that COE can allocate exhaustible resources efficiently because the competition for such resources among current users is likely to deplete the stocks of those resources before future users have the chance to compete for them.[1] This counterargument would seem to be relevant to any resource that is not automatically replenished by natural processes, and it implies that COE should not be allowed to determine the allocation of potentially exhaustible resources.

This reasoning is not without a certain plausibility. Moreover the counterargument holds a powerful attraction for those who question the value of economic growth, believe that modern society has become too materialistic and acquisitive, and believe that we should seek to live as part of nature rather than as masters of it—that nature has an intrinsic value apart from and transcending its usefulness to humans, that we, humans, in short, are not the measure of all things.[2]

It is not my intention here to challenge the moral and philosophical case for making our society less acquisitive, slowing down its rate of economic advancement, or preserving the natural environment. But what I shall contend is that such concerns are groundless insofar as they are based on a presumption that natural resources are going to be depleted under COE. Although there are to be sure environmental problems that have a strong claim on our attention, the exhaustion of resources allocated by COE is not among them. Indeed, it is precisely in cases where resources are not allocated by COE that the problem of exhaustion presents itself.

In this chapter I shall begin by presenting an argument for why depletion is not a problem under COE. In doing so, I use a number of examples showing how COE leads automatically to the postponement of the consumption or the use of resources. These examples bring into focus the role of property rights in conserving resources for future use. Finally, having elaborated on the role of property rights in achieving efficient conservation of resources, I shall outline a simple model showing how, under COE, the consumption of a finite amount of oil could be stretched out indefinitely. In explaining the principles governing the efficient intertemporal allocation of resources and the role of COE in bringing an efficient allocation of

1. This in effect was the thesis of the well-known study by Meadows et al., *Limits to Growth*, sponsored by the Club of Rome.

2. For a description of this attitude see Schurr et al., *Energy*, pp. 404–08, especially p. 406. For examples of this attitude see Rifkin, *Entropy*, and Daly, *Steady State*.

these resources about, I am going to abstract from some important problems that arise out of our uncertainty about what the future holds. Discussion of these problems will have to wait until Chapter 3.

COMPETITION BETWEEN PRESENT AND FUTURE USES: THE CASE OF SPECULATION

Future generations of resource users seem to suffer from a crippling—one is tempted to say fatal—disadvantage in competing for a potentially exhaustible resource with those who would use it now: They are not here. Thus only the demands of current users are registered, and the demands of potential future users are ignored, or so it would seem. If so, it would seem necessary to limit the use of exhaustible resources in the present so that they are not prematurely depleted by current competition for their use.[3]

Appearances, however, are deceptive. While it is true that future generations are not able to compete directly for the use of resources, that does not necessarily mean that no one has an incentive to compete for resources on their behalf. Indeed, under COE a thoroughly selfish motive exists for competing for the use of exhaustible resources on behalf of future generations.

To see why, it will be helpful to first consider another example in which time and expectations about the future play an important role—speculation. Although speculation is one of the most characteristic features of a market economy, professional speculators are not greatly admired—ranking somewhere between used-car salesmen and members of Congress in public esteem. Speculators, for instance, are frequently blamed for wild swings in commodity prices which it is thought they have somehow contrived so as to profit at the expense of both consumers and producers.

But how does a speculator make a profit? Just like anyone else under COE; he must buy low and sell high. Now it is obvious that had the speculator not bought low, the price would have fallen even lower; had he not sold high, the price would have risen even higher. So notwithstanding common beliefs to the contrary, speculators reduce price fluctuations and stabilize consumption over time—pro-

3. This was the point of Meadows, *Limits to Growth.*

vided, that is, that they really do buy low and sell high. If they don't and suffer losses rather than make profits, they destabilize prices. But even then, their losses partially compensate the rest of the community for destabilizing the market.

The point is that even though speculators themselves care about nothing but making a profit, there are those besides the speculators who benefit from their activities. I refer specifically to consumers in the future, who—since speculators withhold supplies in the present for resale in the future—will pay lower prices than they would have had to otherwise.

Consider what would happen after the harvest of a storable crop if there were no speculators. Immediately after the harvest, a relative abundance of the crop exists. If it were all made available to final consumers right away, the market price would be greatly depressed for a short while. At the low price, consumers would quickly exhaust the supply, driving up the price to a very high level until the next harvest, when the price would again fall sharply. Anticipating this price trend, speculators buy up supplies when the price is low and gradually sell off their stocks before the next harvest. By doing so, they make a profit while incidentally serving the interests of those who want to consume between the harvests and who otherwise would either have had to go without the commodity, buy it at a much higher price, or buy the commodity low and store it themselves at a greater cost than they incur when buying from the speculators.

THE ROLE OF PROPERTY RIGHTS

In purchasing now with a view to future resale, speculators serve the interests of future consumers who are more amply supplied with commodities than they would have been without the speculators. Such speculation is only possible when property rights have certain characteristics. For speculation to be carried out, it must be possible to establish and maintain ownership over a commodity or a resource without actually putting the commodity or resource to any current use, and it must be possible to transfer ownership at low cost.

The idea that ownership—which includes the right to exclude all potential users from access to the resource—can be exercised over a commodity or a resource even if the owner does not put it to any current use is a characteristic of private property that may seem not only useless but positively harmful. Why, after all, should someone who is not using a resource at all keep all those who would like to use it from doing so? A notion of private property this strong also evokes hostility because it allows profits to be earned merely through an exchange of property rights without any input of labor by the owner. But despite the hostility they arouse, the rights to exercise and transfer ownership over a resource that the owner is not using is a crucial characteristic of private property without which many of the benefits that flow from that institution could not be derived.

This hostility toward property rights is directed with particular intensity against the almost universally scorned activity of land speculation. Land speculators are blamed for driving up land prices, keeping land idle, and making huge profits. But speculators only realize a profit if they correctly anticipate that some future change in the neighborhood of the land they purchase will raise the value of the land. Speculators withhold the land from current uses that would conflict with anticipated future uses and, thereby, keep it in readiness for those anticipated uses. (If there were current uses of the land that did not conflict with the speculator's anticipated future uses, speculators would be foregoing profits by withholding the land from those uses as well as the conflicting ones.) Of course speculators may be mistaken in their anticipations. If so, they withhold land from current uses that are more valuable than those for which it is withheld. But then they also have to bear the loss implicit in the withdrawal of the land from current usage.

If control over land could only be maintained while it was being used and control over harvested crops could only be maintained by processing or consuming them, it would be impossible to withhold land or crops from current uses in anticipation of more valuable uses in the future. Thus the ability of individuals to compete on behalf of future users with those who would use resources in the present depends absolutely on the enforcement of property rights over resources that can be established and transferred without putting the resources to any current use. Given such a concept of private property, COE allows potential users of a resource in the future to compete for it in the present.

Let me pause here to anticipate and dispose of a possible objection to this contention: We can understand, it might be said, why someone would withhold a commodity on behalf of consumers who, she expects, will pay her higher prices later than consumers are paying now. But given normal life expectancy, who would withhold a resource for consumers two or three generations in the future? No matter how much future generations would have wanted the resources to be withheld, they cannot very well reward someone after her death for having done so.

The inability of future generations to compensate directly those who now withhold on their behalf is only a problem if all generations in the interim lack the foresight we currently have. If they can foresee the demand of subsequent generations as well as we can, then they will eventually reward those who withhold now in the expectation that they, in turn, will be rewarded themselves by still later generations. None of this need be done for altruistic reasons. It would be done simply because of the expectation that the withheld resource would appreciate in value. If this seems farfetched, explain in some other way how it is that wine and cognac made more than a century ago are still being aged for future consumption.

PROPERTY RIGHTS AND THE CONSERVATION OF NATURAL RESOURCES

In this section I want to discuss the role of private property rights in conserving natural resources.[4] I propose to do so by considering the following question: Why are some species threatened by the human demand to consume their flesh, furs, skins, or other body parts, while other species are not similarly threatened by such a demand? To put the question more concretely: Why in the last century did the American bison nearly become extinct, while other bovine mammals flourished? Why today is the leopard endangered and not the mink? Why is the eagle endangered and not the turkey?

The first thing to notice in pondering these questions is that a purely biological explanation will not do. A sufficiently large demand by humans to consume the body parts of any species, however prolific, could ultimately overwhelm its reproductive capacity. But the species in greatest danger are not necessarily those for which human demand relative to reproductive capacity is greatest. The

4. This discussion and much of what follows is based on Alchian and Demsetz, "Property Rights." See also Anderson, *Water Rights*; Deacon and Johnson, *Forestlands*; Libecap, *Locking Up the Range*; and Stroup and Baden, *Natural Resources.*

demand for poultry, I should imagine, exceeds the demand for eagles by at least as much as the reproductive capacity of the former exceeds the reproductive capacity of the latter. Yet the eagle, not the turkey, is the endangered species.

Thus we have to recognize that extinction is as much an economic as a biological and ecological phenomenon. This is not to say that extinction is merely the result of the pursuit of profits. Those who slaughtered bison a century ago were no more eager for profits than those who slaughter cattle today. It is the context in which the desire for profit is allowed to manifest itself that determines whether a species is liable to extinction.

What we have to be concerned with, therefore, are the implications of alternative ways of assigning and acquiring property rights to animals. It is a safe bet that if a species is threatened with extinction, members of that species are common property so that everyone has as much, or as little, claim to them as anyone else. The problem is that when a valuable resource is subject to communal ownership, everyone has an incentive to convert the communal rights into private rights. The communal rights to a whale, for example, can most easily be converted into private rights by killing the whale and taking possession of the body. The incentive to do so is directly related to the value of the dead whale and is weighed against the cost of hunting and killing whales. Whenever the value of the privately held rights increases, say because the products derived from whales become more valuable, the incentive to convert communal rights into private rights increases correspondingly.

What hunters overlook is that if the whale is still growing and it can still procreate, it may be more valuable alive than dead. But they disregard the value of the living whale (and hence part of the true cost of killing it) because to delay killing the whale simply allows someone else to appropriate the whale by killing it first. Thus competition arises between individuals who seek to kill whales before others do so without regard to the value the whales have while alive. An increase in the demand to consume the products that can be derived from dead whales necessarily increases the rate at which whales are killed. If this rate exceeds a critical level, the stock of whales will eventually be depleted,

But this is certainly not the way to maximize wealth. An increase in the demand to consume whales that raises the value of dead whales also increases the value of live ones that can increase the entire stock of whales. Why then doesn't COE ensure in this case, as it is supposed to, that wealth is maximized?

The answer is that only because live whales are regarded as common property while dead ones can be claimed as private property does COE fail to ensure wealth maximization. If you doubt this, think of what happens when the demand to consume beef increases. First the price of beef rises. The value of cattle then rises as well. When the value of cattle rises, cattle raisers respond not by depleting their herds but by investing in larger ones. So an increase in the demand to consume beef in fact leads to an increase in the stock of cattle.

The role of private ownership in conserving wildlife seems to have been well understood more than two centuries ago by the Montague Indians who inhabited large parts of Quebec at the time of the introduction of the fur trade.[5] Until that time, all the land occupied by the Montagues had been held in common. Any game on their land was also common property and could be hunted without restriction. Demand before the advent of the fur trade was probably quite low, so it made little sense for the Montagues to go to the trouble of assigning and enforcing private property rights over the game just to eliminate the modest amount of overhunting that was taking place under communal ownership. This changed with the introduction of the fur trade, when the value of the game and therefore the cost of overhunting increased sharply. But as long as the game was subject to communal rather than private ownership, the incentive to hunt game rose just as sharply. To avoid the problem of overhunting, the Montagues divided up their common land holdings and assigned ownership of game to the owner of the land on which it was found. As long as game did not migrate over long distances, private ownership of land forced each landowner to bear the full cost of killing live game since doing so reduced the stock of game on the owner's land. Because of the subsequent conversion to private ownership of land and game, the introduction of the fur trade into Quebec did not lead to a depletion of the stock of game as it probably would have otherwise.

Thus when private property rights over animals can be established and enforced at relatively low cost, as is true for cattle, sheep, mink, and poultry, their owners can determine the optimal (wealth-maxi-

5. The example of the Montagues is presented in the classic paper by Demsetz, "Theory of Property Rights."

mizing) division of the stock between current consumption on the one hand and investment for the future on the other. But when animals such as bison in the nineteenth century and whales, leopards, and eagles today are common property, they frequently become candidates for extinction.

It is easy—and emotionally satisfying—to blame the threat to survival of these and other species on greed, especially someone else's. But to do so avoids the real issue. Just about everyone is greedy. Whenever it is possible to convert communal rights into valuable private rights for less than the value of the private rights, the communal rights tend to disappear. If communal rights to animals can be converted into private rights by branding each animal, the conversion can take place without disappearance of the animals themselves. Indeed, the stock is likely to flourish. But if the conversion can only be made by killing the animals, their prospects for longevity are not good.[6]

Why aren't there private property rights to living members of endangered species? One reason is that such rights are very costly to establish and enforce, often owing to the fact that the species is migratory. It is too costly to enforce a claim to a private right over a live eagle not in captivity to make asserting such a right worthwhile. Sometimes, however, such rights could be enforced but aren't because the government will not recognize a claim to such a right. In the nineteenth century bison roamed freely on federally owned land. Since the government permitted hunting but would not sell the land on which the bison roamed, the bison could only be appropriated to private ownership by slaughtering them.

6. Sometimes the existence of communal rights in live animals that can only be converted into private rights by killing the animals has particularly shocking consequences. Each year hunters go to Prince Edward Island in the Gulf of Saint Lawrence to slaughter baby seals. The seals, which are valued for their skins, are publicly owned, and the Canadian government sets an absolute limit on the number that can be killed so as to prevent depletion of the stock. Owing to the limit on the total number of seals that may be killed, each hunter has an incentive to kill and skin the seals as quickly as possible in order to maximize his take before the total legal limit is reached. The seals are brutally slaughtered and even skinned alive as the hunters rush to obtain as many skins as they can. What is responsible for the outrage is not simply the greed and ruthlessness of the hunters but a procedure for assigning rights to the seals that makes speed the primary condition for success in hunting them. Since humane methods of killing are slower than inhumane ones, the overall limit in conjunction with communal ownership encourages brutality. See Alchian and Demsetz, *Property Rights*, p. 20.

To sum up: The tendency toward overexploitation of a resource by current users at the expense of future users (of which extinction of a species is a special case) is a consequence of a system of mixed communal and private property rights that allows the communal rights to be converted to private rights by subjecting the resource to a current use that is incompatible with future use. The problem of overexploitation in the present can be overcome when private property rights to the resource can be established without first devoting the resource to a current use. But if it is too costly to establish and enforce private property rights to the resource without first subjecting it to a current use, some other method of restricting the competition to convert communal rights into private rights may be required.

NONRENEWABLE RESOURCES

I have shown how private property rights contribute to the conservation of natural resources. But you may note that there is a difference between the kinds of resources discussed in the previous section, all renewable, and resources like oil that are not. One might think that even private property rights cannot ensure the efficient husbanding of nonrenewable resources, which it would appear must ultimately be depleted. But our discussion should have established that the recognition of private property rights over a nonrenewable resource like oil while it is still in the ground implies a slower rate of extraction than if private property rights over the resource are recognized only after it has been extracted.

Just why this is so and how certain other factors affect the rate of extraction can be illustrated with a simple example.[7] Think of two children, Billy and Bobby, sipping soda through a straw. Consider two cases. Suppose that in one case Billy and Bobby are sipping from the same glass and in the other that each sips from his own glass. If the children have the same amount of soda to drink in both cases, in which one will they drink the soda more quickly?

It is obvious that they will drink more slowly in the second case. In the first case Billy knows that any soda he doesn't drink now is available for Bobby to drink. Each tries to drink faster than the other

7. I owe this example to Harold Demsetz.

because the one who drinks the fastest gets more soda. But in the second case, both know that any soda not consumed now will still be there later. Each can drink as quickly or as slowly as he pleases without worrying about how fast the other is drinking.

How fast each chooses to drink in the second case would, of course, be affected systematically by several factors. For example, if either child were afraid that someone else might take his glass away before he finished drinking, he would drink faster than otherwise. The rate at which they drink will similarly be affected by the rate at which the soda spoils (loses carbonation). And the more soda the children expect to obtain in the future, the faster they are likely to drink in the present.

This example illustrates how private ownership (the two-glass case) slows down the rate of consumption of a nonrenewable resource relative to the case of common ownership (the one-glass case). It also suggests how expectations about the future influence the current rate of consumption under private ownership. With this as an introduction, I will describe a more formal model of how private owners would go about depleting a given stock of a nonrenewable resource.

HOW TO MAKE OIL LAST FOREVER

Suppose that there were no more oil to be found anywhere in the world so that when the last barrel of oil was extracted from current reserves that, let us say, were known with certainty, the stock of oil would be totally depleted. Future generations would only be able to use oil saved from currently existing reserves. How would the existing reserves of oil be allocated among potential users in the present and future under COE? And what sort of allocation of these reserves would be efficient?

A cautionary note is in order. Although many discussions of resource conservation are conducted as if this assumption were true, it is not at all descriptive of reality.[8] New reserves of oil as well as

8. The literature on the economics of natural resources in general and exhaustible and nonrenewable resources in particular is rapidly growing. A classic discussion of the entire subject is Scott, *Natural Resources.* A recent work that has incorporated more of the current technical literature and therefore operates at a higher level of abstraction and mathematical sophistication is Dasgupta and Heal, *Economic Theory and Exhaustible Resources.* A

new substitutes for oil are constantly being found and are likely to continue to be found for a long time to come, so our future is not nearly so bleak as this assumption suggests. Nevertheless, there are two good reasons for making this assumption anyway. First, it brings out nicely the way in which expectations of future conditions affect current decisions about how to allocate resources between current and future uses. Second, the assumption represents a sort of worst-case scenario. If COE can handle the worst case scenario, that at least suggests it can also handle other less bleak scenarios.

One more assumption I want to make before proceeding with the analysis is that there are no direct costs of extracting oil. In other words, oil producers can extract whatever oil they want to without incurring any sacrifice in the process of extraction. I have two reasons for making this assumption. One is that it will simplify the subsequent analysis without detracting from the essential logic of the argument. The other is that this assumption is also, in a sense, a worst-case scenario (from a conservationist standpoint) because it means that there is no physical barrier to the immediate and total depletion of the stock of oil.[9] This assumption will thus be very useful in highlighting the incentives for conservation under private as opposed to communal ownership.

With these preliminaries taken care of, I will proceed to the analysis. A private owner of oil reserves has three basic choices: (1) to extract all reserves now and sell at the current market price, (2) to leave all reserves in the ground now and extract later, or (3) to extract some reserves now and the rest later.

If there were no costs of extraction, what would keep the owner from extracting and selling all the reserves immediately? The way to

---

recent textbook by Griffin and Steel, *Energy Economics and Policy*, is also quite useful. An excellent discussion of the problem of exhaustible resources can be found in Williams, "Running Out." Penetrating discussions on the economics of the question can also be found in Gordon, "Economics and the Conservation Question," and Simon, *The Ultimate Resource.* The source for much of the later work in the field of exhaustible resources is Hotelling, "The Economics of Exhaustible Resources." See also Stroup and Baden, *Natural Resources.*

9. From another and more relevant standpoint, zero extraction cost is a best-case assumption because it implies that there is no sacrifice that has to be made in order to consume oil. To the conservationist this appears to be a worst-case assumption because it makes current consumption easier, and the object of the conservationist is to postpone consumption. The goal of policy should not be to postpone current consumption but to optimize it, that is, to maximize the benefits from consumption over time. To do so requires selecting an optimal time path for consumption. Such a time path obviously does not involve perpetual deferral of consumption just because consumption depletes the remaining stock.

approach this question is to recognize that keeping oil in the ground is simply one way of investing. The question is then, What circumstances make investing in oil reserves profitable? The answer is that such investment is profitable when oil appreciates at a rate exceeding the rate of interest. If it does not, selling oil now and collecting interest on the proceeds would generate a larger return than investing in oil in the ground.

One may wonder why oil should be expected to appreciate, but in fact appreciation is inevitable given the assumption of a fixed nonrenewable stock of oil. Selling more oil now means that less will be sold in the future. Since selling more oil now lowers the current price, while selling less in the future will raise the future price; the more oil that is sold now, the more rapidly oil must appreciate in the future. If such a small amount of oil were being sold now that the implied rate of appreciation were less than the rate of interest, owners would have insufficient incentive to invest in oil. Not investing, however, means selling the oil now so that the current price would necessarily be forced down, and the implied rate of appreciation increased until the incentive for renewed investment were restored. If so much oil were sold in the present that the implied rate of appreciation were above the rate of interest, owners would increase their investment in oil by reducing their current sales. The reduction in current sales would drive up the current price and reduce the implied rate of appreciation until the incentive to sell in the present was restored.[10]

10. For this argument to be strictly valid, there must either be correct foresight on the part of all transactors or there must be a complete array of forward markets in which contracts for delivery at future dates can be entered into. The condition that the rate of appreciation of oil equal the rate of interest would determine the structure of the spot and forward prices at every moment. In practice there is not perfect foresight, and there is only a limited set of forward and futures markets. But the same forces that would operate with complete forward markets still operate under limited futures markets. Current decisions are still made in light of expected future conditions and those who can anticipate future conditions most accurately will in fact earn the largest profits. Moreover, institutions like the stock market are at least partial substitutes for the absent futures markets. If there were a general expectation that oil would become much more valuable in periods further in the future than contracts for future delivery are being made, there is still the opportunity of buying up reserves in the ground. If the value of reserves in the ground rise relative to the current price, that indicates a market expectation of higher future prices. The market value of firms with relatively large reserve holdings compared to current sales would increase relative to firms with relatively small holdings of reserves. This would signal firms to try to build up their reserve position as one way of increasing their stock value. One way in which this could be done, of course, would be to slow down current extraction rates, which would force up the current market price.

Thus, under COE, our assumptions determine an equilibrium rate of extraction. This rate implies a rate of increase in the price of oil just equal to the rate of interest. Although our intuition may rebel at the suggestion, such a rate of extraction does imply that current reserves can be made to last forever (or at least until the world comes to an end or until the price rises to the level at which the demand for oil vanishes). For as long as further appreciation at the rate of interest is anticipated, private owners will continue to withhold oil as investment for the future.

The condition that the price of oil appreciate at a rate equal to the rate of interest does not of course fully determine the price path because it may start rising from either a low or a high level. Two factors that would help to determine the initial price are the size of the stock of oil and the demand over time. The greater the stock of oil, the greater the amount of oil that can be consumed over time and, thus, the lower the initial price. Similarly, the lower demand initially and the lower the rate of demand growth, the lower the initial price. (See Appendix A.)

Some may regard the appreciation of the existing stock of oil as undeserved windfall that ought not to accrue to private owners of oil, but consider what would happen if oil reserves, instead of being privately owned, were communally owned. Since any oil not extracted immediately by one producer could be extracted the next moment by another, everyone would have an incentive to convert the communal rights to oil beneath the ground into private rights to oil above ground. When a private owner extracts oil, even if she bears no direct cost of extraction, she must still bear what is called a "user cost." The user cost represents the revenue foregone by not extracting the oil at a later date. A communal owner, however, bears no such cost because the choice is not between extracting now and extracting later but between extracting now and not extracting at all. Thus communal ownership would result in current extraction being increased until the price of oil equaled the direct cost of extraction. Given our assumptions, this means that the price would be forced down to zero. Extraction would continue at this rate until the entire stock were depleted, at which point the price would rise to infinity (or whatever price would be necessary to reduce the amount demanded to zero).[11]

11. Of course, if oil could be privately owned and stored above ground, the price of oil to consumers would not go to zero, and the entire stock of oil would not be depleted right

Since private owners of oil must bear the user cost of current extraction, they will extract a barrel of oil in the present if and only if its current value is at least as great as the future value of a barrel of oil discounted by the rate of interest. Thus the allocation of oil reserves between current and futures uses under private ownership and COE maximizes the value of those reserves. A barrel of oil is extracted in the present if, and only if, its current value is at least as great as the future value of a barrel of oil discounted by the rate of interest. It would therefore not be possible to increase wealth by either increasing or decreasing the current rate of extraction.

On the other hand, it is easy to see that under communal ownership wealth could be increased by decreasing the rate of extraction since the current value of oil would be far less than the future value of oil discounted by the rate of interest. Although investing in oil underground would be socially very profitable, no one would have the incentive to undertake it if oil reserves were communally owned.

CONCLUSION

I have shown in this chapter how private property rights force all current users to bear the full cost of using any resource, which cost corresponds to the most valuable use, either present or future, that must be sacrificed because of the current use. The owner of a resource maximizes wealth by putting the resource to its most valuable use or else does not cover the costs of ownership. But if a current use conflicted with a future use that is more valuable, the owner, or someone else who buys the resource, would reserve the resource for that future use.

I showed how this sort of intertemporal allocation takes place in instances such as commodity and land speculation, the husbanding of animal species, and the conservation of oil reserves. It was evident that not only are exclusive and transferable property rights essential to the achievement of an efficient intertemporal allocation of resources but also that the process through which private property rights can be established is itself critical to the efficiency of resource

away because there would be an incentive to buy up oil now to store for future use. But since it is much more costly to store oil above ground than to leave it underground, the lack of private property rights to oil reserves would still increase current consumption at the expense of future consumption.

allocation. Unfortunately, the establishment or enforcement of property rights to certain resources can sometimes be achieved only by exercising physical control over the resource, and sometimes this can only be done by putting the resource to a current use (e.g., consuming it), which conflicts with potentially more valuable future uses.

The basic idea is that any process for acquiring private property rights over resources besides COE involves a waste of whatever resources are used up in the process of acquiring those rights. Thus whenever people compete with each other to claim the property rights to a resource, the resources expended in the process are wasted. But it should be understood that the waste arises not because of private property and COE but because of their absence. When for technical reasons private property rights cannot be instituted over certain kinds of resources, we may have no choice except to adopt some method of competition other than COE. Nevertheless the difficulty in assigning private property rights can sometimes be overcome by advances in technology (such as surveying, which made it easier to enforce private property rights in land) or by new institutional arrangements that allow the establishment of private property rights over resources that were previously held in common.

In discussing the problems of intertemporal competition and intertemporal allocation of resources, I have so far avoided another kind of problem: our inevitable ignorance of what the future is going to be like and the possible uses that will be found for resources in the future. I have done so in order to concentrate on the logic of the intertemporal allocation process under private ownership and COE. Having explained that logic, I must concede that in a world of pervasive uncertainty about the future, the logic of intertemporal allocation is not nearly so straightforward as it seems in the tidy world I have been discussing where the future can be foreseen with reasonable certainty.

Neither the amount of oil available now nor the amount that will be found in the future nor the kinds of substitutes for oil that will be discovered in the future are in any definite sense known to us. Some may suggest that without such knowledge the case for private property and COE is undermined. Moreover, uncertainty about the future frequently gives rise to an extreme pessimism or conservatism about nature and our environment. Such attitudes have permeated a great deal of recent conservationist thinking that opposes the use of private property and COE as mechanisms for managing our resources and that, in its more extreme forms, disputes the very idea of human progress.

In the next chapter I shall try to explain why our uncertainty about the future does not undermine the case for private property and COE and why it does not require the kind of extreme pessimism about our use of natural resources and the environment that some conservationists have expressed.

Chapter 3

HOW TO CONSUME OIL WHEN YOU DON'T KNOW HOW MUCH IS LEFT

The conservationist who urges us to "make greater provision for the future" is in fact urging a *lesser* provision for posterity.

Anthony Scott*

INTRODUCTION

In the previous chapter I showed how, under an appropriate legal framework, COE could indefinitely extend the use of nonrenewable resources. In doing so, I used an intertemporal model that assumed the ultimately available supply of oil as well as all future demands for oil were known in advance. Although the model is useful in drawing attention to the nature of market forces that balance the potential future uses of any resource against its current uses, the relevance of the conclusions derived from the model to actual conditions may seem questionable because of its very strong assumptions about what we know of the future.

Since decisions concerning the rate at which to use up natural resources are more complicated than the model used, I shall now address the decision problem in a more realistic setting. But we

*The epigraph to this chapter is taken from Scott, *Natural Resources*, p. 97.

should not forget that in one sense the assumptions made in the previous chapter were extremely unfavorable to future generations in that they excluded the possibility either that unknown resources or inventions could replace currently known supplies. Nevertheless, we found that COE could extend the useful life of a finite amount of the resource for an indefinite period.

It cannot be denied, however, that we do incur a risk in consuming a nonrenewable resource because we do not and cannot know how much of the resource or what kinds of substitutes for it will be found in the future. If future discoveries of oil turn out to be meager, our consumption now may turn out to have deprived future generations of what may be a vital resource for them. Thus under such conditions of uncertainty and incomplete knowledge, the risk that excessive current consumption could ultimately cause the exhaustion of future resources appears more serious than the analysis of the preceeding chapter suggested. A very cautious policy towards nonrenewable natural resources in order to ensure that future generations will not be deprived of a share in the earth's limited endowment of natural resources would seem to be called for.

My purpose in this chapter is to address this conservationist argument. I contend that the argument overlooks the risks to the welfare of future generations in consuming natural resources too slowly now. Given both kinds of risks, it is not obvious that we can improve very much on the rate of consumption generated by the market. After dealing with the conservationist argument in its more traditional form, I shall examine a new, more radical, conservationist position that, according to its exponents, is a necessary result of the second law of thermodynamics, also known as the entropy law. I propose to show that this conservationist position does not follow either from the laws of thermodynamics or those of economics.

TWO KINDS OF ERROR IN THE USE OF NATURAL RESOURCES

If indeed we do not know how much oil there is left to be found and if indeed we do not know how great future demands for oil are going to be, it may well appear that it is not only prudent but imperative to reduce current consumption of oil and other nonrenewable resources as much as possible. Otherwise our overuse of natural resources may actually jeopardize the well-being, if not the survival, of future generations. This in brief is the conservationist case for restricting the consumption of natural resources.[1]

It is easy to see why excessive consumption of nonrenewable resources is an error for which posterity would have to suffer. But it is equally true, though less obvious, that too little consumption of nonrenewable resources is an error for which future generations—not just our own—would also have to suffer. On first hearing, the notion that increasing our consumption of nonrenewable resources could be beneficial to future generations may sound paradoxical, but the paradox is easily dispelled. One simply has to recognize that future generations are going to benefit from the man-made resources—machines, factories, buildings, roads, bridges, and so on—as well as from the natural resources they inherit from us. If by depleting the stock of nonrenewable resources more rapidly we can expand the stock of manufactured resources, future generations may well gain as a consequence.

WHAT SHOULD WE BEQUEATH TO POSTERITY?

Since decisions about how rapidly to deplete the stock of natural resources are subject to both of the errors just mentioned, the conservationist argument for not using up limited natural resources implicitly assumes that natural resources will be more valuable to posterity than will the resources we make of them. To see how questionable this assumption is, just consider that we inherited fewer natural

1. Since I am going to be rather critical of certain conservationist positions in this chapter, I should say that I do not reject all conservationist positions and objectives. Conservationists act legitimately in seeking to prevent common property resources of all types from being overused. But this is not the same as insisting that those resources should remain government owned. It is also legitimate for them to expreas their *preferences* for natural as opposed to manufactured resources and to promote their values and preferences by political action and other appropriate means. It is less acceptable, however, for them to assert that their values are morally superior to differing values or that their agenda is somehow ordained by the findings of science. For further discussions of conservation and conservationism, see Scott, *Natural Resources*, Barnett and Morse, *Scarcity and Growth*, and Simon, *Ultimate Resource*. Also see Gordon, "Economics and Conservation," pp. 110-21, Hayek, *Constitution*, pp. 367-75, Alchian and Allen, *University Economics*, pp. 426-28, Kay and Mirrlees, *Economics of Depletion*, pp. 140-76, and Stroup and Baden, *Natural Resources*.

resources and more manufactured resources from our parents than our grandparents did from their parents. Nevertheless, we are certainly materially better off than our grandparents were. In fact it is largely because our parents and grandparents did use up natural resources that they were able to create the stock of capital that now helps to support us at a much higher standard of living than our parents and grandparents enjoyed.

Furthermore, if we restrict the term "resource" to something that is known about by, and is useful to, at least one person, it may well be that we inherited a larger stock of natural resources from our parents than our grandparents did from theirs. This may also sound paradoxical, especially if I add that this only came about because our parents and grandparents used up natural resources as fast as they did. The paradox is dispelled as we recognize that by finding more uses to which natural resources could be put and thus increasing the value of those resources, our parents and grandparents created new incentives for finding previously unknown and undreamt of supplies of those resources. Had no one used the resources for fear of depleting them before future generations could use them, much of the stock of resources that we know about today would never have been discovered.

A century ago few people knew or cared about crude oil, and natural gas was just a nuisance that frequently had to be put up with in the process of extracting crude oil from the ground. Today, crude oil is the primary fuel source for much of the world, and natural gas is so valuable that millions of dollars are spent in efforts to find it even if it is not associated with crude oil.

The best we can do for posterity is not to leave them with the largest possible stock of natural resources but to leave them with the stock of manufactured, as well as natural resources that has the greatest value. Nevertheless, we do not necessarily want to *maximize* the amount of wealth we bequeath to our offspring because increasing the size of our bequest to our children requires us to reduce our own consumption. We care for our children and do reduce our consumption for their sake, but not without limit. At some point an increment in our bequest is no longer worth the further sacrifice of our own consumption that it would require.

How much consumption to forego for the sake of our children is not the same question as in what particular form to leave the wealth we accumulate to our offspring. A decision about how much consumption to forego for the sake of our children is a savings decision; a decision about whether to convert a natural resource into some other form is an investment decision. Though conceptually distinct, the two kinds of decisions are interrelated. Greater opportunities for investment imply increased incentives for saving, while a greater desire to save will make more investment profitable.

An example will clarify these distinctions. Suppose I own a tree, and let us say that I have only four options with respect to the tree: (1) I can cut it down and use it for firewood to keep warm or to cook with; (2) I can sell the tree either standing or after I cut it down and put the proceeds in the bank to earn interest; (3) I can cut it down and use it to build something, say a cottage; or (4) I can let it grow. Option (1) is a decision to consume the wood now. Whether I choose (1) or (2) through (4) will depend on my willingness to sacrifice the present enjoyment of the firewood or some other consumption of equal value in order to accumulate greater wealth for myself and my heirs. Options (2) through (4) all involve decisions to save rather than consume now. For example, (2) involves putting my savings at the disposal of someone else who will invest and provide me with a fixed return. If I do (3), I am using my savings to invest in a cottage. In (4) I am using my savings to invest in a tree.

We can oversimplify a bit and look at my decision as a two-stage process. First comes the decision between (1) and (2) through (4). If I choose (1), there is nothing else to decide. If I don't, I must make a further decision among the remaining options. Suppose I do not choose (1); any of the remaining choices involves the sacrifice of current consumption equal to the value of the wood that is now in the tree. What determines my choice among (2) through (4)?

Let us say that there are 1,000 pounds of wood in the tree now and that the wood sells for $1 per pound. If I sell the tree, I shall have $1,000 to put in the bank from which to draw interest. For now, say the interest rate is 10 percent annually. If I decide to cut down the tree and use the wood to build a cottage, (3) I shall be sacrificing $1,000 of current consumption in return for the future services that the cottage will provide me and my children or whomever it is sold to. For simplicity I am going to assume that the cottage will last forever, and let us assign a market value of $100 per year to the housing services provided by the cottage.

If I do (4), I shall be sacrificing $1,000 worth of current consumption for a greater amount of wood in the future. Suppose that in one year the tree will grow to weigh 1,200 pounds, and say that the value of wood will remain $1 per pound in the future. After a year I may cut down the tree, or I may reinvest in it by letting it continue to grow.

Now to decide whether to invest as well as to save and, if so, which investment to undertake, I must discount future values at some rate in order to calculate their present value equivalents. Since the rate of interest represents my alternative use of the funds available for investment, the interest rate is the obvious candidate for a rate of discount. I have assumed that the interest rate is 10 percent per year. If so, the present discounted value of the $1,200 of wood I can obtain one year hence by investing in my tree is about $1,091. Thus investing in the tree is clearly better than simply saving $1,000 at 10 percent interest. Is it better than investing in a house? If I invest in the house I shall be acquiring a perpetual stream of services that have a value equivalent to an infinite sequence of annual payments of $100. If discounted at a 10 percent interest rate, the present value of this stream is worth exactly $1,000. (If you do not see why this is so, imagine that you had $1,000 in the bank earning 10 percent interest. You could withdraw $100 at the end of each year forever. Hence the equivalence of the present sum and the perpetual sequence of future payments.) Thus investing in the tree is more profitable than investing in a cottage or just saving at a 10 percent interest rate.

Now suppose the interest rate is 5 percent. Investing in the tree is more profitable than it was at 10 percent. The present value of $1,000 in one year is about $1,143. But investing in the cottage becomes even more profitable than investing in the tree because the present value of $100 per year forever at 5 percent is $2,000. Thus at a 5 percent interest rate, investing in the cottage becomes the most profitable alternative.

But at a 25 percent interest rate, investing in either the cottage or the tree would be unprofitable. The value of the tree next year ($1,200) would be equivalent to only $960 today, while the present value of the cottage would only be $400. The upshot of this is that the interest rate affects both the amount and the form of investment.

As long as future values are estimated with some accuracy, the sorts of investments that are made will tend to make the future better off than if other investments had been made instead. Thus, at a 5 percent interest rate, investing in a cottage makes a greater contribution to future well-being than investing in a tree; and if I must choose between the two, I shall invest in the cottage. This choice is not only in my interest, it is in the interest of future generations since the future return that I am counting on is going to be derived from the services the cottage is going to provide to future occupants.

Thus insofar as it will benefit future generations that we eschew the current use of natural resources, there are corresponding incentives in the marketplace for investment in those resources (provided that private property rights in these resources are defined and enforced so that, for example, I know that if I do not cut down my tree today someone else will not come along tomorrow and cut it down instead). If some of these resources are being depleted, it means that, given the current discount rate of the future relative to the present, it is more profitable to convert some of those natural resources into manufactured form than to keep them in their original state.

One could restate the point as follows. If trees grow fast enough, it might not be worth it to cut them down now to convert them into houses because our children would gain more from having such an immense supply of wood. But if trees don't grow all that fast (or if the value of trees would fall sharply if the supply of wood expanded significantly), then we may help our children more by cutting down trees and building houses and bridges with them so that our children will have the use of those synthetic resources already available to them instead of having to build them.

If you doubt this assertion, just ask yourself whether we should be richer or poorer today if previous generations had not converted natural resources into manufactured form. The answer, it seems to me, is that we should undoubtedly be poorer. Indeed, if no such transformations had ever been performed, we would still be living in hunting and gathering tribes.

If underprovision for future generations is a problem, then the solution lies in reducing the rate of discount used to evaluate investment projects. The lower the rate of discount, the greater the weight attached to future sums in present-value calculations. If you want to increase provisions for the future, find a way to reduce market interest rates. One way to do so would be to exempt interest income from taxation; another would be to terminate the tax deductibility of interest payments.

Since it would make investments of all kinds more profitable, a reduction in the interest rate would increase the rate of capital accumulation and thereby raise the standard of living future generations could attain. Although it might appear that reducing the interest rate would also reduce the rate at which nonrenewable resources were depleted, as the model of Chapter 2 suggested, this tendency could be offset by the interrelationship between the interest rate and the demand for nonrenewable resources. At a reduced interest rate, it might become profitable to undertake investments that were relatively intensive in their use of nonrenewable resources such as oil.[2] If that were the case, reducing the rate of interest might actually speed up the depletion of nonrenewable resources.

HOW CAN FINITE RESOURCES SUSTAIN A GROWING POPULATION AND RISING LIVING STANDARDS OVER TIME?

In the previous section I suggested that conservationist opposition to transforming natural into manufactured resources may be contrary to the interests of future generations. A conservationist response to this might be that although using up natural resources can temporarily allow population size and living standards to increase, it cannot do so indefinitely. Eventually natural resource depletion must have seriously harmful, if not disastrous, effects on future generations. It therefore appears reckless and perhaps ultimately suicidal to use up the earth's natural resources at a rate that cannot be sustained indefinitely.

I don't think that this response in any way undermines my argument. Let me explain why not.

It is certainly true that we cannot continue to use up the earth's resources at a rate that cannot be sustained indefinitely. The problem is to determine what rate is sustainable. But even if the current rate is not sustainable, it is not necessarily a mistake to use up resources temporarily at an unsustainable rate. Depletion need not result because so long as there are effective property rights over resources, the prospect that any resource for which there were not good substitutes is going to be depleted would cause the price increases needed to discourage consumption. For resources that are common property, of course, special conservation measures may be appropriate.

2. Scott, *Natural Resources*, pp. 95–96.

The problem with identifying a priori a sustainable rate of resource use is illustrated by history: Conservationists 200 years ago thought the rate of use then was not sustainable, and conservationists 100 years ago thought that the rate of use then was also not sustainable. That earlier conservationists were wrong, however, does not prove that today's conservationists are wrong too, but it does raise the suspicion that the conservationist view of the world systematically underestimates the earth's capacity to support life and to support it at rising living standards.

The core of the difficulty is that conservationists view the world as if it were a completely closed system. That is implicit in their concern about the finiteness of resources. Although there are, indeed, limits on this world and the resources available, the limits are not so rigid as conservationists imagine. We can act in various ways to make those limits less confining. We live in an open world and even an open universe. At a minimum, our world is open to the discovery of new knowledge. This new knowledge, in turn, helps to relax limits that would otherwise have constrained our growth.

To William Stanley Jevons, it seemed that the coal reserves of 100 years ago placed an effective limit on the growth and development of the British economy.[3] Long before coal even approached exhaustion, however, it was displaced as the world's primary fuel by petroleum. It seems fantastic to imagine, but it is equally possible that some new discovery will eventually render oil relatively unimportant or obsolete as an energy source.

It is simply not possible to know now what resources future generations are going to be most in need of. The ability of future generations to provide for themselves is going to depend on knowledge that no one today knows or perhaps even imagines. Those living in the last century had equally little conception of today's knowledge because to know now what will only be known later is clearly contradictory.[4]

If we don't know what future generations will know—among other things, what resources they will demand and what supplies they will have discovered—it is impossible to specify beforehand an efficient, or even sustainable, rate of resource use over time. To do so

3. Jevons, *The Coal Question.*

4. This has been shown by Karl Popper, *Poverty of Historicism*, pp. vii–viii. A formal proof is now available in Popper, *Open Universe*, pp. 68–77.

would require that future knowledge be known in advance. All efficiency statements made by economists presume that knowledge is already given, if not to a single mind, than at least in dispersed form to all agents within the model. But if the knowledge in the model cannot even be specified, the notion of efficiency—as normally used by economists—breaks down. (This is no criticism of either economists or the concept of efficiency; it is just a recognition of the limits to a particular intellectual construct.) In fact we live in a world in which not only is knowledge growing in an unpredictable fashion, but also the particular courses of its growth will depend on the way in which resources are used in the present. The science of geophysics, for example, would be far less advanced today if there had been no demand for crude oil or if geophysical knowledge had not been useful in finding new supplies of crude oil.

To use up resources only to the extent that our present knowledge of current supplies suggests they are available would actually deprive us and posterity of the use of resources that are now, or could eventually become, available to us. F. A. Hayek has summed up this point well:

> We are constantly using up resources on the basis of the mere probability that our knowledge of available resources will increase indefinitely—and this knowledge does increase in part because we are using up what is available at such a fast rate. Indeed, if we are to make full use of the available resources, we must act on the assumption that it will continue to increase, even if some of our particular expectations are bound to be disappointed. Industrial development would have been greatly retarded if sixty or eighty years ago the warning of the conservationists about the threatened exhaustion of the supply of coal had been heeded; and the internal combustion engine would never have revolutionized transport if its use had been limited to the known supplies of oil. . . . Though it is important that on all these matters the opinion of the experts about the physical facts should be heard, the result in most instances would have been very detrimental if they had the power to enforce their views on policy.[5]

The best we can hope for is (1) to make decisions about using particular resources that are based as much as possible on the available knowledge concerning the following: (a) currently available supplies, (b) current substitutes, (c) potential future supplies, (d) potential future substitutes, (e) current uses, and (f) potential future uses; and

5. Hayek, *Constitution*, pp. 369–70.

(2) to stimulate the growth of such knowledge. Perhaps the chief justification for the market is precisely that it does make full use of available knowledge and does promote the discovery of useful new knowledge.[6] Information about current or future resource supplies and demands is always affecting market prices and altering patterns of use in ways appropriate to the new information. Credible projections that future supplies are going to be tighter than previously anticipated immediately increase current prices, reduce current consumption, and provide increased incentives for the discovery of both new sources of supply and new substitutes.

We cannot be certain that the market is right. In fact, since the market price is always a distillation of conflicting opinions, it probably is not. But the sensitivity of the market to all available information implies that the market price is an unbiased predictor of future conditions. It is as likely to be underoptimistic as it is overoptimistic with respect to the future supply of resources.[7] Anyone who disagrees with the anticipations upon which the current price of any resource and the current rate of its depletion are contingent is free to buy up supplies now to withhold for future use. Should her expectations turn out to be correct, the speculator will become very wealthy and, by adding to the relatively inadequate stock of the resource in the future, will have performed an important public service. The tendency of price controls or a windfall profits tax on crude oil, natural gas, or other nonrenewable resources to discourage just such speculative stockpiling is one of the strongest arguments that can be raised against them.

Even if anticipations about the supplies of a particular resource turn out to be overoptimistic, it is a certainty that this would be realized before the resource was actually depleted. As expectations were corrected, upward revisions in the price would restrict usage to prevent depletion before a substitute for the resource was found. Even if a particular resource were completely exhausted because its future

6. See Hayek, "Use of Knowledge," reprinted in Hayek, *Individualism*, pp. 77–91, and Hayek, *New Studies*, pp. 179–90.

7. In Appendix A, I present an argument that the market may be biased towards too much conservation of a nonrenewable resource. If its initial price were high enough, a resource could appreciate at a rate equal to the interest rate and a portion of the stock would never be consumed. Market forces would not necessarily force a price reduction. But if the initial price were too low, market forces would eventually raise the price of the resource to avoid depletion.

supplies had been continually overestimated, it cannot seriously be suggested that this would happen to all resources simultaneously. But once one resource was mistakenly exhausted, the market would increase its estimate of the probability that other resources would be exhausted. Resource prices would rise, and their rates of usage would fall as a consequence.

That resource prices have generally been declining over time suggests that, so far at least, we have not been depleting natural resources. Indeed, the growth of knowledge—which, as Julian Simon argues so forcefully, is a function of population size[8]—creates new resources for us by enabling us to discover previously unknown supplies of known resources and to discover ways of using materials that we previously did not know how to use. Nothing guarantees that this trend will continue, but neither is there any compelling reason to believe that the process is being or soon will be reversed. Even if it were reversed, our best hope for managing scarcity would still be to let market forces allocate the remaining stocks, thus providing human ingenuity with incentives to use the still-available resources more sparingly.

The advantage of relying on COE and the market rather than on the judgment of any particular individual, however knowledgeable, or even on the judgment of a particular group of experts is that the market continually subjects different opinions to the test of experience. It rewards correct opinions and penalizes incorrect ones while encouraging everyone to adjust his actions to his own special information and to the general information that is reflected in market prices.

Perhaps the best that can be said for the conservationist is that he is trying to find what in game theory is called a "min-max" strategy. That is to say, we can view nature as presenting us with various potential future states that each have a certain probability of arising. States with very large endowments of resources offer the opportunity for large gains that can be exploited by an appropriate strategy, and states with very small resource endowments result in very large losses that may be avoided by an equally appropriate strategy. The conservationist can be thought of as trying to find whatever strategy will minimize the loss that would be borne should the worst possible state of nature be realized. But when the payoffs from all possible

8. Simon, *Ultimate Resource.*

states of nature are weighted by their probabilities, the expected payoff implied by the min-max strategy may be much smaller than the payoff implied by a less conservative strategy.

Although a min-max strategy is not an irrational one to follow, few of us adopt it in our daily lives. Most of us obviously do accept slightly higher risks of some undesirable, or even catastrophic, event when the expected benefits are sufficiently attractive. Otherwise we should not be traveling in cars or airplanes, climbing up mountains or skiing down them. One could argue that it makes more sense to adopt a min-max strategy for the world as a whole than for any single individual to do, but even so, it is not clear that conservationism is a true min-max strategy, especially if the market does sooner or later establish a price for any resource that reflects its "true" scarcity. And the greater the risk associated with depleting any resource, the more likely it is that the market price of that resource would prevent its depletion.

Nor should we forget that there are catastrophes to which humanity has been subject over the years that are different than the ones conservationists are preoccupied with. For millenia people were imperiled by famine and epidemic. The black death reduced the population of Europe by one third in the fourteenth century. The growth of knowledge, productivity, and population have been essential in freeing an ever-growing proportion of the world's population from those catastrophic dangers.[9] There can be little doubt, therefore, that in the past conservationism would not have been a min-max strategy.

THE ENTROPY LAW AND ITS MEANING

The idea that conservation is a min-max strategy emerges from a number of writers who have sought to derive a conservationist program from one of the fundamental laws of nature, the second law of thermodynamics, also known as the law of entropy. The concept of entropy is, roughly, that the universe as a whole is becoming increasingly disorderly and chaotic. Ultimately, therefore, all resources must become degraded and unusable, so the hope of indefinitely continuing to find more resources with which to support a growing popu-

9. Simon, *Ultimate Resource.*

lation at a rising standard of living is futile. Building on the work of a learned and world-renowned economist, Nicholas Georgescu-Roegen—who unfortunately does not seem to have grasped the full implications of the economic theory of property rights and market processes—these writers, whom I shall call entropists, have been advocating fundamental changes in our political, social, and economic institutions designed to reduce our dependence on energy sources that *necessarily* must be exhausted.

An important example of their thinking can be found in a recent book by Jeremy Rifkin called *Entropy: A New World View.*[10] This book, which includes a laudatory afterword by Professor Georgescu-Roegen, has received considerable notice in many of the country's leading journals of opinion, and it provides a theoretical justification for many of the most extreme views that have been gaining currency within the environmental movement: the evils of population and economic growth, the exploitation of the resources of the underdeveloped world by the developed world, the arrogance of our attempts to control or alter the natural environment, the futility of seeking solutions to our problems in technological and scientific progress. Rifkin argues vehemently not only that economic growth is undesirable but also that it is rapidly becoming impossible, that we are about to enter a new era in which the high-technology, energy-intensive civilization we have built on the basis of ample supplies of inexpensive fossil fuels is going to be unsustainable. Our only choice is to adopt less specialized, less energy-intensive modes of production and ways of living.

Some of these ideas may be attractive for their own sake (nor would I discourage people from experimenting in different ways of living and producing), but the burden of Rifkin's argument seems to me antithetical to human values in its virtual denial of human creativity and cultural and intellectual advancement. To Rifkin and his fellow entropists, the entropy law means that the universe and everything in it is moving towards degradation.[11] Thus anything that contradicts this universal tendency towards degradation, as do human creativity and intellectual advancement, must be dismissed as either an illusion or reinterpreted as a manifestation of the process of degradation. So I believe it is important to show that such attitudes do not follow, as Rifkin asserts, from the entropy law.

10. Rifkin, *Entropy.*
11. Rifkin, *Entropy*, p. 46.

Before proceeding, I must first briefly explain what the entropy law says and what it means, though I make no claims to expertise in physics or thermodynamics. The ideas I shall be discussing, however, are so basic that it seems to me clear thinking is more important than expertise in the subject. My concern is not so much with the law of entropy as with applying it sensibly to real problems.

The second law of thermodynamics can be formulated in a number of essentially equivalent ways. Let us start with the following formulation. Matter and energy (which, according to the first law of thermodynamics, must be conserved so that there can be no increase in the amount of matter or energy in the universe) tend to move from more to less orderly states. For example, if a gas is held within a container, the walls of the container will keep the molecules of gas in a relatively orderly arrangement. But if the container is opened, the gas will not continue to stay within the container, that is, remain in an orderly state. Instead, the gas molecules will escape and distribute themselves randomly through space, and the gas will ultimately be dissipated in the surrounding environment. Before the container was opened, the gas was in a structured and relatively orderly arrangement; after the container was opened, the gas moved into a completely unstructured, disorderly, random arrangement. The same tendency, moreover, is operating – though at a much slower rate – on the container itself since the container will, after sufficient time, also assume a less orderly structure, gradually eroding and eventually disintegrating. No loss of matter occurs as erosion takes place, but the bits of matter gradually become less and less structured.

Similarly there can be no loss of energy, but energy can either be available or unavailable, free or bound. The entropy law tells us that available energy tends to become unavailable. By available energy, I mean energy that is available for, or free to perform, physical work – to move mass. In order for it to be free and available, energy has to be in an orderly state, for example, a chunk of coal or a barrel of crude oil. In such a state it can be burned to generate heat. If the heat generated can be concentrated in a small enough space, the tendency of heat to flow from a warmer to a colder body (another way of formulating the second law) can be marshaled to perform work. The heat generated creates a temperature difference that can be exploited. But once the heat has been dissipated, though it has not been lost, a temperature difference no longer remains to be exploited; no further work can be performed. At that point the energy is unavailable or bound.

This explains why the second law can be formulated as a denial of the possibility of a perpetual motion machine—a machine that will operate indefinitely after an initial application of energy. Free energy propels a machine by creating greater heat in one part of the machine than in the rest of the machine. As heat is dissipated within the machine and lost to the surrounding environment, the temperature difference must, according to the second law, be eliminated. Once there is no longer a temperature difference that can be exploited, the machine requires a further injection of free energy to continue to operate, so perpetual motion is not possible.

Entropy is a measure of the total disorder in any physical system. The second law of thermodynamics can thus be reformulated as follows: In the universe as a whole or in any closed physical system within the universe (such as our solar system) the level of entropy is increasing. This does not mean that order cannot be increased; it just means that free energy must be expended to do so. I can arrange a given collection of materials in a more orderly fashion and (at least temporarily) increase the orderliness of those materials by, say, making a house out of them. But to do so, I will have to expend a certain amount of free energy that will no longer be available. The increase in entropy associated with the expenditure of energy will necessarily be greater than the reduction of entropy associated with the construction of the house. This is no cause for alarm (although some entropists think it is) if the services the house can provide are valuable enough to me or to others who would pay me for it. That entropy has increased because I build the house does not mean the world is then a worse place to live. Whatever free energy I have made unavailable would either be made unavailable later by someone else or would never be used by anyone, in which case it would have been rendered unavailable by purely physical processes. The only relevant question is whether my use of the energy was more valuable than any other potential use.

Another question may have occurred to you concerning the compatibility of the entropy law with evolutionary theory. Evolutionary theory asserts that increasingly complex forms of life have come into being over the course of time. Evolution seems to imply, contrary to the entropy law, that order has been increasing and entropy decreasing rather than the other way around.

Is there a fundamental conflict between evolutionary theory and the entropy law? Not really because the entropy law only holds for a physically closed system. But evolution takes place in an open system—the earth, which is open since it receives free energy, or sunlight, from the sun. The flow of solar energy has made it possible for life on earth to increase in quantity and to achieve increasing degrees of complexity and order. Not only has biological order been enhanced, but inanimate order has also increased as the result of the work done by living things, from the honeycombs of bees and the nests of birds to the dams of beavers and the bridges of humans.[12]

Of course, the first and second law require that the books be balanced. So the increasing order on earth has been paid for, as it were, by decreasing order in the sun. As it generates energy and heat, the sun is slowly burning itself out, slowly dissipating into the void. Luckily for us this burning-out process will go on for millions of years before it has any noticeable effect on terrestrial life. Note also that the use life on earth has made of the sun's energy has had no adverse effect on the sun. By enjoying the sun's warmth and light, we do not make the sun work any harder or burn out any faster.

There is one other source of free energy that, in principle, is almost as fertile as the sun. This source arises because it is possible to convert matter into energy at a phenomenally high rate of transformation given Einstein's famous equation $E = MC^2$, where E stands for energy, M for mass, and C for the speed with which light travels. Thus from a small amount of matter, enormous amounts of free energy can be obtained. Unfortunately, the very productivity of this kind of transformation makes it enormously dangerous and costly to undertake.

I do not feel competent to render any judgment about the merits of nuclear energy. The crux of the issue, however, is whether we are willing to incur the costs required to make nuclear energy production acceptably safe. There seems to me no doubt that acceptable safety is technically possible. What is not so clear is how great the costs of achieving acceptable safety really are. Without explicit or implicit subsidies to nuclear power, it might not be possible to provide it as cheaply as energy from conventional sources. Improving knowledge and technical advancements, though, are likely to increase the safety and reduce the cost of providing nuclear energy as time passes. We may decide never to exploit it, but if we decide not to, it will be be-

12. F. A. Hayek has provided further explanation of the possibility of evolution of more complex structures despite the law of entropy. See Hayek, *Fatal Conceit*, Chapter 1.

cause its costs relative to other energy sources will remain too great to make nuclear energy worthwhile. Nevertheless, the possibility of deriving energy from nuclear reactions means that at a sufficient cost, we shall be able to obtain large amounts of energy for an indefinitely long period. This places an upper limit on the costs we shall have to incur in order to derive energy from other sources.

MISUNDERSTANDING THE ENTROPY LAW

It can be laid down as a general rule that every attempt to derive a sweeping historical trend, explanation, or law that is supposed to predict the future course of history or social evolution—whether it be Hegel's dialectical idealism, Marx's dialectical materialism, or Georgescu-Roegen's entropism—will be falsified by future events.[13] In Georgescu-Roegen's case, however, the falsification is a bit more tricky since everyone acknowledges that the world and the universe must ultimately come to an end. It is far from clear what possible observation could falsify Georgescu-Roegen's predictions since it could always be maintained that the prediction would be realized at some later date.

Nevertheless, it is clear that the ultimate principle of the universe, according to Georgescu-Roegen, is degradation. Ever since the first seconds of creation, the universe has been decaying. Although life and its evolution to higher forms appear to violate the principles of degradation, in reality life merely speeds up the process of degradation. The interpretation bears some similarity to classical Greek views that all change represented decay or degradation.[14]

Let us consider the way Jeremy Rifkin formulates the entropy principle and the implications he draws from it:

> The overall disorder of the world is always increasing; the amount of available energy is always decreasing. Since human survival depends upon available energy, this must mean that human life is always becoming harder and harder to sustain and that more work, not less, is necessary in order to eke out an existence from a more and more stingy environment. Because there is not enough time in a day for human beings alone to perform the additional work

13. Popper, *Poverty of Historicism.*
14. Rifkin, *Entropy*, pp. 10–12 and Popper, *Open Universe*, pp. 11–17.

> required by the harsher energy environments, more complex technologies must be devised . . . just to maintain a moderate level of human existence.[15]

Lest you think that improving technologies offer a way to escape from the harsh reality dictated by the inexorable increases of entropy, Rifkin makes the following observation:

> We also entertain the belief that technology is creating greater order in the world, when again, the opposite is the case. The entropy law tells us that every time available energy is used up, it creates greater disorder somewhere in the surrounding environment. The massive flow-through of energy in modern industrial society is creating massive disorder in the world we live in. The faster we streamline our technology, the faster we speed up the transforming process, the faster available energy is dissipated, the more disorder mounts.[16]

Thus, according to Rifkin and other entropists, we are running a losing race to overcome the increasing disorder entailed by the entropy law, and the harder we try to keep up, the faster we fall behind.

Rifkin builds his case upon two fallacies. One is that all increases in entropy in our environment are harmful to us. In fact Rifkin explicitly equates pollution with higher entropy. Now all pollution may be associated with higher entropy, but not all increases in entropy are polluting. Whether an increase in entropy causes pollution depends on how people are affected by it and how people react to those effects. When water is at the top of a waterfall, it contains available energy because it can run a turbine as it falls. After it reaches bottom, the energy is still there but is no longer available. Thus entropy has increased, but it would be absurd to associate this increase in entropy with pollution.

Sometimes the effects of an increase in entropy may even be pleasurable. Suppose you are driving your car on a very cold day. The engine of your car converts fuel into heat in order to move the car, but in the process the engine also emits heat, that is, unavailable, bound energy, high entropy. Yet the heat your car's engine emits helps to keep you warm. Aside from accomplishing work, the increase in entropy has also improved your environment, not polluted it.

15. Rifkin, *Entropy*, p. 66.
16. Rifkin, *Entropy*, p. 80.

Rifkin's second fallacy is even worse. The overall disorder of the world is not always increasing. It would only be increasing if the world had no outside source of energy, but the earth absorbs and stores a constant flow of energy emitted by the sun. As we use up some of the stored solar energy, entropy does increase, but that entropy increase will occur anyway. If oil were left in the ground indefinitely, the bound energy there would eventually be dissipated when the world is burned up by the sun in two billion years. The only question is how we can make good use of a state of energy that ultimately will decay regardless of whether we use it or not.

Rifkin is led to similarly misguided conclusions when he attempts to explain evolution in light of the entropy law.

> We are so used to thinking of biological evolution in terms of progress. Now we find that each higher species in the evolutionary chain transforms greater amounts of energy from a usable to an unusable state. In the process of evolution, each succeeding species is more complex and thus better equipped as a transformer of available energy. What is really difficult to accept, however, is the realization that the higher the species in the chain, the greater the energy flow-through and the greater the disorder created in the overall environment.
>
> The entropy law says that evolution dissipates the overall available energy for life on this planet. Our concept is the exact opposite. We believe that evolution somehow magically creates greater overall value and order on earth. Now that the environment we live in is becoming so dissipated and disordered that it is apparent to the naked eye, we are for the first time beginning to have second thoughts about our views on evolution, progress, and the creation of things of material value.[17]

This is exactly backwards. Evolution has generated increasingly complex forms of life that are capable of utilizing more of the free flow of solar energy and more of the already accumulated stock of solar energy contained in all life forms and in fossil fuels than are less complex forms. The alternative to using the free flow of solar energy is not a reduction in entropy. The increase in entropy is occurring on the sun and is not affected by what happens on earth. Using up stored energy whether in biomass or fossil fuels does imply an increase in entropy. But this increase, as I showed above, does not necessarily involve any deleterious effects on the environment. It merely represents a kind of accounting transaction required by the first law

17. Rifkin, *Entropy*, pp. 55-58.

to show that energy has been transformed from an available to an unavailable state. The increase in entropy is a measure of this transformation, but does not necessarily have any further environmental implication.

The availability of solar energy, either sunlight or stored up, allows the effective amount of order on earth to increase—if not indefinitely, at least for enormously long periods of time—provided we use the energy wisely. And it is the growth of scientific knowledge and the advance of technology that enables us to make increasingly effective use of the energy.

Thus the entropy law does not dictate that things, as Rifkin insists, are always getting worse despite superficial appearances. What the entropy law does teach us is that all available sources of energy are, whatever we do, eventually going to become unavailable. It therefore behooves us to try to discover new sources of available energy and new ways of using the known sources so that we derive the maximum use from our available, but wasting, energy endowment.

More disturbing than faulty conclusions about the capacity of our world to support population growth and rising living standards is the entropists' hostility towards the idea of human progress—material, intellectual, and especially scientific. Such hostility, of course, is by no means confined to entropists. It is increasingly widespread throughout our society, particularly among extreme environmentalists. The danger is that entropists purport to provide a sort of weirdly ironic scientific justification for this hostility.

Let us again turn to Rifkin for a forceful statement of the values he believes are implicit in the entropy law.

> Whenever energy is extracted from the environment and processed through society, part of it becomes dissipated or wasted at every stage, until all of it, including that which is made into products, ends up in one form or another as waste at the end of the line.
>
> Most economists simply can't buy this simple truth. They are wedded to the idea that human labor added to nature's resources creates greater value, not less. Because machine capital is ultimately viewed as past labor mixed with resources, it too is considered to be creating economic value. They can't get it into their heads that machines and people *can't create anything. They can only transform the existing available energy supply from a usable to a wasted state*, providing only "temporary utility" along the way.[18]
>
> (My emphasis.)

18. Rifkin, *Entropy*, p. 129.

Thus in one breath, Rifkin dismisses the music of Mozart and Beethoven, the plays of Molière and Shakespeare, the novels of Austen and Tolstoy, the paintings of Rembrandt and Monet, and every other creation of the human mind and spirit. Not only does he deny the reality and even the possibility of human creativity, he holds every human work to be inherently destructive. Consider:

> The more energy is available, the greater the prospects for extending the possibilities of life into the future. But the second law also tells us that the available store of energy is continually being depleted by *every occurrence. The more energy each of us uses up, the less is available for all life that comes after us. The ultimate* [sic] *moral imperative, then, is to waste as little energy as possible.*[19] (My emphasis.)

The self-contradictory nature of Rifkin's position can be seen here quite clearly. Available energy, Rifkin concedes, does enhance the prospects for life, but it cannot contribute to the enhancement of life except by being used and converted to an unavailable state. Rifkin has, however, already *defined* this conversion to be a waste since it is associated with an increase in entropy. His moral imperative of wasting as little energy as possible is thus tantamount to a moral imperative to use as little energy as possible or, in other words, to leave as much energy as possible unused. This of course must ultimately mean minimizing the "prospects for extending life into the future."

It could perhaps be said in Rifkin's defense that he was merely exaggerating the dangers of using energy resources in order to encourage their conservation for future use. We have already seen, however, that subjecting natural resources to private ownership and allowing market prices to determine their rate of use over time improves the chances for making reasonably good use of those resources and reducing the risk of their depletion so long as no substitute resources are found to take their place.

Rifkin, however, explicitly denies that markets can conserve nonrenewable resources for future uses:

> It should be understood that there is no way to allow for the needs of future generations in classical economic theory. When we meet as buyers and sellers in the marketplace, we make decisions based on the relative abundance or scarcity of things as they affect us. No one speaks for future generations at

19. Rifkin, *Entropy*, p. 255.

> the marketplace, and for this reason, everyone who comes after us starts off much poorer than we did in terms of nature's remaining endowment. Imagine what it would be like if all future generations for the next 100,000 years could somehow bid for the oil our generation is using up. Obviously, the price of that energy would be so expensive that it would be prohibitive if future generations were allowed [sic] to participate in today's resource allocation decisions.[20]

It would be hard to imagine a greater misunderstanding of the role of private ownership in bringing to bear the interests of future generations on current decisions about the use of resources than that displayed in this passage.

Another entropist, Herman Daly, has tried to show that although market prices are effective means of allocation in the face of what he calls "relative scarcity," they cannot handle the problem he calls "absolute scarcity."

> Correctly adjusted relative prices allow us to bear the burden of absolute scarcity in the least uncomfortable manner. But even an efficiently borne burden can eventually become too heavy. When the relative price of the relatively scarce resource rises, as it eventually will, it induces the substitution of relatively abundant resources. Price cannot deal with absolute scarcity because it is impossible to raise the relative prices of all resources in general. Any attempt to do so merely raises the absolute price level, and instead of substitution (What substitute is there for resources in general, low entropy?), we merely get inflation. Maybe that is one of the root causes of inflation in advanced economies.[21]

The distinction between relative and absolute scarcity that Daly offers would not be tenable even if we accepted that sources of low entropy were, as he suggests, already being depleted. The various sources of low entropy about which Daly is worried—coal, crude oil, natural gas, and so on—are not all running out at the same rates, so substitution among relatively more and relatively less scarce resources is indeed possible and takes place all the time. Moreover, sources of low entropy, like sunlight, are not for all practical purposes running out, and substitution will take place as other sources of low entropy do run out.

It is also rather embarrassing to find that a professionally trained economist such as Professor Daly would suggest that the depletion of

20. Rifkin, *Entropy*, p. 137.
21. Daly, *Steady State*, p. 42.

sources of low entropy was responsible for inflation. Even if this were the case, depletion of some natural resources would not cause the prices of *all* resources to rise. It would only cause the prices of those that were being depleted to rise. In addition to solar resources, there are still other resources, notably capital and labor, whose relative prices would fall. The prices of labor- and capital-intensive products would fall to offset the increase in the prices of products intensive in the depleted resources. Nor can the relative price of everything rise in an inflation. The relative price (i.e., the purchasing power) of money necessarily falls. If the relative price of money falls, that indicates an excess supply of money has been driving down its relative price. Daly's statement that the depletion of low entropy may be responsible for inflation might have had some surface plausibility if inflation had been occurring with a constant money supply. But since money supplies (however measured) have been increasing rapidly along with inflation, it is simply ludicrous to attribute inflation to a depletion of the world's stock of low entropy.[22]

CONCLUSION

It is natural to worry about what is going to become of ourselves, our children, and our children's children. Our concern for our future welfare and that of our offspring induces us to make provisions for the future by saving, investing, and accumulating wealth. In doing so we tend to leave the world as a whole a wealthier place when we go than when we entered it. The progress of civilization confirms that people generally have been concerned with the welfare of succeeding generations. Indeed, such concern is arguably the strongest motive for the economic development of humanity.

There is, however, another kind of concern for future generations that seems to have emerged in the two centuries since Malthus wrote his *Essay on Population.* The new concern has not been so much a solicitude for particular individuals people care about as it is a theo-

22. I have discussed elsewhere (Glasner, "Relative Prices") the fallacy that increases in the cost of widely used resources can be held responsible for inflation, which is actually a phenomenon related to the supply of and demand for money. By also attributing inflation to the ability of noncompetitive or monopolistic industries to pass on their higher resource costs to consumers, Daly (*Steady State*, pp. 42–45) also commits the other fallacy I addressed in that paper.

retical anxiety for humanity in general inspired by the belief that the resources of the world will be unable to support a population that becomes too numerous or too prosperous. The goal of those motivated by the new theoretical concern for the future thus becomes almost the opposite of the older instinctive concern for the future. The goal becomes to prevent humanity from growing either too numerous or too wealthy. This goal is, of course, intended for the ultimate good of mankind because it is feared that though humanity might become very numerous and perhaps very wealthy for a short time, in the process of so doing they would exhaust the world's means of support.

For a variety of reasons such conservationist arguments have been found unconvincing by most economists. The main reason is that the argument only makes sense if people act as if the resources they were using were not scarce. But the whole purpose of private property rights and market prices is to ensure that people do recognize the limits of scarcity in their use of resources. Moreover, these institutions guarantee that future conditions, insofar as they can be foreseen, are taken into account by those currently using resources. More important, there is a constant effort to find new and less costly ways of doing things because such limits restrain everyone. Thus new knowledge is continually being sought out, discovered, and brought to bear to relieve scarcities where they become most pressing.

Conservationists have sought to bolster their case for limiting the growth of wealth and population with an argument inspired by the second law of thermodynamics, the entropy law. This argument asserts that since the entropy law tells us that the universe is becoming more disorderly and degraded all the time and since the only way to reverse the process within a given part of the environment is by application of free energy, our use of free energy now only speeds up the process by which entropy reaches its maximum. As Georgescu-Roegen has put it with disarming frankness, "If we stampede over details, we can say that every baby born now means one human life less in the future. But also every Cadillac produced at any time means fewer lives in the future."[23]

What the argument fails to recognize is that the quantity of low entropy available for exploitation has not been given to us in advance. We must learn how to use it, and we must find it before it

23. Georgescu-Roegen, *Entropy Law*, p. 304.

decays on its own. This quest requires human knowledge, and it requires incentives both for the use of that knowledge and for the discovery of new knowledge. Part of the process that leads to the growth of knowledge is the birth of babies today who will add to our stock of knowledge in the future; another part is the production of Cadillacs, which are among the incentives that promote both the use and the discovery of knowledge. Generations in the future are unlikely to benefit if we impede that process today.

PART II

POLICIES, PRACTICES, AND EVIDENCE

Chapter 4

THE EVOLUTION AND STRUCTURE OF THE GASOLINE MARKET

INTRODUCTION

Perhaps no political, social, or economic issue occupied so much public attention during the 1970s as did the energy crisis and what to do about it. Undoubtedly much of that urgency stemmed from two brief, but traumatic, periods when life in many parts of the United States was totally disrupted by gasoline shortages. These shortages were widely thought to be both manifestations of an energy crisis and precursors of even more serious disruptions that—in the absence of a comprehensive national energy plan—would surely not be long in coming.

Nevertheless, the belief that those shortages had anything to do with an actual energy crisis seems to me almost entirely without foundation. I have already suggested that the real reason for those shortages was that price controls during both of these episodes restricted the terms of trade that could be offered and accepted by buyers and sellers of gasoline. There were controls on gasoline during both of the episodes in which shortages occurred, but the controls were in force from August 1971 to January 1981, while shortages were only noticeable from the fall of 1973 to the spring of 1974 and again in the spring and summer of 1979. So one could easily conclude that the controls did not really have much to do with the shortages and that something else must have been responsible. Indeed, during neither episode did controls figure very prominently among competing explanations for the shortages that in one way or another related them to an energy crisis.

Yet none of the competing explanations for the shortages – the Arab oil embargo, the fall of the Shah, the decline in U.S. oil production, or some conspiracy on the part of the oil companies – makes any sense as a cause of shortages. At most such explanations could account for increased prices, but certainly not for shortages. There is just no plausible explanation for these shortages besides price controls.

However for price controls to qualify as a valid explanation of the gasoline shortages, the fact that for nearly nine of the nine and a half years that gasoline prices were controlled no shortages occurred must also be explained. Moreover, if price controls were for the most part not resulting in shortages, were the controls merely ineffectual and superfluous during those periods or were they having effects that were manifested otherwise than through shortages?

To understand these problems in any depth, we shall have to get acquainted with some background information about the organization of the gasoline market and the specifics of the controls over gasoline prices. In this chapter I will discuss the evolution and the structure of the gasoline market, and in the next two chapters, I will give an account of the imposition of price controls and their subsequent application.

We must begin with a description of the complex network of intermediaries between the refiners and the final users of gasoline. Without some feel for how the network operates, it is impossible to understand either the controls on the retail market or the response of the various market stages to the controls. To get such a feel, it will help to consider how the market for gasoline and the distribution network evolved. It soon becomes apparent that the structure of the market is not necessarily arbitrary or inefficient but is rather the outcome of a continuing process of adjustment and adaptation to the preferences of consumers and to the relative costs of alternative methods of supplying them with gasoline.

Of course in any discussion of the structure of the gasoline market the role of the major oil companies – called "the majors" – is a matter of intense interest. The role of the majors in marketing gasoline and the extent of vertical integration exhibited by them have long given rise to allegations that the majors use vertical integration to facilitate predatory tactics against their nonintegrated counterparts in the industry. So I will also consider the economic rationale for direct marketing of gasoline and for vertical integration by the major oil companies. We shall find that charges of predatory competition do not hold up very well in the light of such analysis.

THE STRUCTURE OF THE GASOLINE MARKET

Whenever we speak of the market for this or the market for that we are speaking metaphorically. We use the term "market" not to describe a particular place where buyers and sellers meet to trade with each other but as a metaphor for a process that encompasses far more than just the physical locations at which those trades take place.

The Gasoline Market and Its Submarkets

The market for gasoline in the broad metaphorical sense consists of an almost innumerable number of narrower submarkets at the refiner's gate, at terminals, at bulk plants, and at ports. There are wholesale markets and there are retail markets. There are spot markets at which sales take place between anonymous buyers and sellers, and there are contract markets in which buyers and sellers are directly linked in more or less long-term relationships. These markets are, however, all interrelated, which is what entitles us to speak of "the gasoline market."

In the absence of constraints, transactors can switch from one market to another as changing conditions alter the terms available in the various markets. Because of the opportunities for substitution, changing conditions are rapidly communicated to all transactors even if the changes were initially perceived by only a handful of them. Thus if unusually large quantities of imported gasoline became available on the Gulf Coast, some buyers would switch from their customary suppliers to take advantage of reduced prices in Gulf Coast markets. To avoid accumulating excess inventories, sellers whose customers had switched to the spot markets on the Gulf Coast would probably reduce their prices to attract new customers—perhaps inland marketers for whom Gulf Coast supplies are too costly even at depressed prices. A lengthy sequence of further adjustments would ensue as the entire network of interrelationships among buyers and sellers rearranged itself in response to the initial change in conditions. Thus when we speak of *the* gasoline market, we mean this entire network of relationships among buyers and sellers of gasoline.

I shall refer to these interrelated submarkets in the next two chapters as I try to explain how the gasoline market worked under controls. For now, however, let us now turn our attention to the different functional stages of the gasoline market.

Refining

By any measure gasoline is the most important product refined from crude oil. In volume, a bit more than 40 percent of refinery output is gasoline. Middle distillates, consisting of heating oils and diesel fuel, account for about 17 percent of refinery output. Residual oil makes up about 13 percent of refinery output, and jet fuel comprises another 6 percent of output. The remainder consists of petrochemical feedstocks, kerosene, ethane, liquefied petroleum gases, asphalt, road oil, and miscellaneous products.[1]

The relative yields of the different products depend on the characteristics of the crude oil used in the refinery and on the refining process itself. The lower the sulfur content and the lighter the weight of the crude, the more gasoline and other light refined products will be produced. Gasoline yields can also be increased by altering the refinery process in various ways. Thus refiners can, within limits, alter the output proportions in response to changes in the relative values of the various products. Of course changes in refinery structure are costly and time consuming, so in the short run the output mix is less flexible than it is in the long run.

As of December 31, 1982, 225 refineries were in operation in the United States.[2] The largest 20 refiners produce over 70 percent of the nation's total output of refined products.[3] Some of the smallest refiners have such limited facilities that they produce no gasoline at all, only heavier products like residual oil. The production of gasoline

1. *Monthly Energy Review*, January 1982, pp. 36-42.
2. *Oil and Gas Journal*, March 31, 1983, p. 130.
3. American Petroleum Institute, *Market Shares*, p. 80.

requires such substantial investments in specialized capital equipment that unless a refinery produces a relatively large amount of gasoline, the capital investment required will usually not be profitable. Thus the share of gasoline in total output tends to rise with the capacity of a refinery.

Because of the economies of scale in the production of gasoline, it might seem that small refineries would be inefficient and, without government subsidy, unprofitable.[4] However, even without subsidy, small refiners can sometimes be competitive when they economize on costs that larger ones must bear, like the costs of transporting crude oil to the refinery and then delivering the products from the refinery to final users. When the retail market is very dense geographically, the savings in production costs accruing to large refineries generally more than offset the additional transportation costs they incur by not being smaller and closer to consumer markets. In the Rocky Mountain and the Great Plains states, however, the large distances separating retail markets of no great size imply such high transportation costs that large refineries are not competitive with smaller ones.

Most of the larger refiners are vertically integrated; they search for, produce, and transport crude oil, and they also have extensive retail networks for marketing under their own brand names. These are the so-called majors, though no undisputed criterion exists for classifying a particular company as a major. Most counts, however, include at least fifteen companies, of which the largest are Exxon, Mobil, Shell, Texaco, Standard of California, and Gulf, six of the fabled seven sisters.[5] Other important majors are AMOCO, Phillips, Union, and ARCO.

That so large a proportion of gasoline is produced by vertically integrated companies has led many to conclude that the gasoline market is largely unresponsive to normal market forces because vertical integration allows oil companies to avoid any market transactions until they sell the gasoline to final consumers.[6] But even though the major integrated companies engage in all stages of the process of producing gasoline, they still undertake upstream transactions—with

4. Small refiners were beneficiaries of the so-called small-refiner bias that was part of both the Mandatory Oil Import Quota Program and the subsequent entitlements program.

5. The seventh sister is British Petroleum, which owns majority interest in Standard Oil of Ohio (Sohio).

6. See for example Allvine and Patterson, *Highway Robbery*, and Blair, *Control of Oil.*

each other and with nonintegrated firms. The mere fact that an integrated refiner produces crude oil does not necessarily make it unprofitable for the refiner to sell some of its crude to other refiners or to buy crude from other sources. It may be more costly for a refiner to transport its own crude to one of its own refineries than to obtain crude for that refinery in the nearby market. An integrated refiner may also trade in the crude market simply because its crude production either exceeds or, more typically, falls short of its refining capacity.[7]

Despite the dominance of the majors, there are also substantial companies—Ashland and Clark are good examples—that primarily engage in refining and marketing; they obtain supplies of crude for their refineries in the market. Such nonintegrated companies are generally referred to as "independents."[8]

Distribution

After gasoline is refined, it must be delivered to locations that are accessible to final users. This might seem a simple enough task, but thousands of firms are required to accomplish it. Independent marketers buy gasoline from refiners and then market the gasoline under their own brand. Many refiners, of course, market gasoline under their own brand names, often also buying gasoline produced by other refiners for resale. Retail operations are sometimes operated directly by the marketers and sometimes by franchised dealers. To supply their franchised dealers, refiners usually sell their supplies to jobbers who in turn sell the supplies to the dealers. Some gasoline is also sold in bulk to final users, such as local governments or car rental agencies that have their own fleets of automobiles.[9]

Refiners sell their gasoline at different points on the way from the refinery gate to the retail pump. At the refinery rack, sales are made to independent marketers with their own retail outlets but with no

7. American Petroleum Institute, *Market Shares*, p. 37.

8. This is an interesting inversion of the normal meaning of the word "independent," since it is presumably the integrated company that is most independent of other companies and nonintegrated companies that are most dependent on other sources for their supplies of either crude oil or refined products. The word "independent" here is not used descriptively but rather because of its emotive value, calling to mind a number of virtues, chief among which, one suspects, is that the companies are not majors.

9. See Fleming, *Gasoline Prices*, and Allvine and Patterson, *Highway Robbery*.

refinery capacity or even to refiner-marketers that require additional supplies for their retail network. Any of the product not sold at the refinery gate is shipped by tanker, barge, or pipeline to the refiner's terminals. From there it can be sold to other marketers, bulk end-use purchasers, and to the refiner's jobbers.

Jobbers are intermediaries who contract with a refiner that has a network of retail outlets operated by franchised dealers to purchase the refiner's output and deliver it to its dealers. Jobbers may own some of the retail outlets they supply, or they may act strictly as intermediaries. The jobber provides the refiner with on-the-spot knowledge of local conditions and personal contacts with dealers and other business people. Such knowledge and contacts would often be more costly for the refiner if it used its own employees to perform the intermediary function.

From the terminal the gasoline may be transported directly by truck to the retail outlet, or it may be transported to an intermediate bulk plant owned either by the refiner-marketer or the jobber. At the bulk plant additional sales can be made to jobbers or to retail outlets for final sale.

Purchase prices, of course, increase to reflect transportation costs as the gasoline flows farther from the refinery gate. Jobbers pay different prices for gasoline depending on where they are buying it and how much of the transportation costs to the retail outlet they must incur themselves. Nor do jobbers generally pay the same price for the product that other customers pay at the same location. Since the jobber is purchasing gasoline to be sold under the brand name of the refiner-marketer, he normally pays a premium to cover the costs of advertising and other services the refiner incurs in his behalf.

At retail, gasoline is sold at a number of different outlets. Most gasoline is sold under a major brand name.[10] The majors market their gasoline either through employee-operated stations that are the property of the company or through franchised dealers. Increasingly both franchised dealers and especially company-owned stations sell some or all of their gasoline at self-service pumps. The franchised dealer providing full service was for many years the typical retail outlet. In the late sixties and seventies, however, the growing popularity of self-service stations along with decreasing demand for routine service and maintenance, along with competition from other suppliers, cut

10. U.S. Department of Energy, *Oil Supply Shortages.*

sharply into the share of the market held by franchised full-service dealers. Independent refiner-marketers have also, though to a lesser extent, made use of franchised dealer networks. Their dealers generally provide minimal service and sell at a discount. Gasoline is also marketed by some department stores (e.g., Sears, K-Mart, and Montgomery Ward), and—in an interesting reversion to one of the earliest marketing methods in the retail gasoline market—many of the small drive-in convenience grocery stores becoming popular in recent years have also begun to market gasoline.

THE EVOLUTION OF THE RETAIL MARKET FOR GASOLINE

The retail market for gasoline is unique among markets for petroleum products. Unlike heating oil, gasoline is generally not delivered to the final user but is obtained by the consumer at the retail outlet. Brand names in the gasoline market pertain to the product, while in the market for heating oil, brand names are specific to the individual retailer and not to the product. Nor are there any nationwide chains of retail outlets selling heating oil comparable to those in the gasoline market.

Such differences suggest that we ought to try to understand the unique characteristics of gasoline that have led to a radically different marketing structure at the retail level than that used to market other petroleum products. The suggestion that the unique structure of the gasoline market has a function other than to exploit consumers will be met with some skepticism by critics of the oil industry;[11] nevertheless, there are economic forces that dictate different marketing practices for gasoline and heating oil. Avarice alone cannot account for the differences, since the oil companies are presumably no less grasping in one market than in another. We can gain some valuable insight into the reasons for the unique structure of the gasoline market if we consider briefly how the market has evolved over the past eighty years or so.[12]

11. For example Allvine and Patterson, *Highway Robbery*, and Blair, *Control of Oil.*

12. The following account of the evolution of the gasoline market relies heavily on Hogarty, "Gasoline Marketing."

Developments Prior to World War II

In the first decade of this century, before automobile ownership became widespread, gasoline was not usually sold by specialized retail outlets. In urban areas gasoline was frequently sold in garages, whose services were in fairly high demand because of the propensity of the early automobiles to break down frequently. For quick purchases motorists could buy gasoline at curbside pumps. These were owned and easily operated by a single individual. In rural areas country stores that sold a wide variety of goods also began to sell gasoline to motorists. Another early method, now used to deliver heating oil, was home delivery of gasoline from mobile trucks, but this quickly failed because of the frequency of gasoline explosions at home storage tanks.

A major difficulty for consumers in those days was the sale of adulterated or low-quality gasoline. The difficulty arose because of the high cost of determining whether the retailer or the refiner was responsible for the low-quality product. Since many customers were, in any event, unlikely to be repeat purchasers, individual retailers had little incentive to provide good-quality gasoline. A retailer could blame the refiner for poor quality, but if heavily reliant on repeat sales, the retailer would seek to ascertain the quality of the gasoline obtained from the refiner in order to guarantee that quality to customers. A low cost way of dealing with this problem was to establish some form of brand name common to numerous retailers of gasoline. The brand name would provide some additional information to the motorist about the quality of the gasoline bought from a branded retailer, specifically, that the retailer was willing to risk that the consumer would shun all the brand's outlets should she be sold defective gasoline at any one of them. This information increases the confidence the consumer can place in the retailer and the incentive of the retailer to provide the quality he claims to be providing.[13]

Before long refiners, who like consumers had suffered from deceptive retailing, began to market gasoline under their own brand names from their own retail outlets. The trend toward integration coincided with the introduction of the first specialized service stations around 1910. These stations evolved from the original curbside pumps and

13. See Klein and Leffler, "Market Forces in Performance."

allowed motorists to obtain gasoline without standing in the street and blocking traffic. They typically had several pumps, which allowed greater station volume than curbside pumps or garages. In addition, they were equipped with large underground storage tanks that economized on the number of deliveries suppliers had to make. Automotive products and minor service were also provided at some of these stations. At first, most stations were owned and operated by refiners using the stations to build up brand names. As early as World War I, though, leasing arrangements under which refiners leased the stations to retailers and agreed to supply products to the lessees became a common, but not yet predominant, form of retailing.

Yet as late as 1927 most retail outlets dispensing gasoline were still either primarily selling other products or services (e.g., garages) or were curbside pumps operated by one person.[14] Most service stations were then owned by independent oil companies with no refining capacity or else by private individuals. The rest were owned and operated by refiners or leased by them to private individuals.[15]

During the thirties two trends helped to make franchise management the dominant mode of service station operation. First, some states began to levy taxes on chain store outlets. This increased the cost of directly operating a station compared to the cost of leasing it to an independent dealer who would use the refiner's products and operate under the refiner's brand name. Second, during the depression workers unable to find opportunities for employment increasingly became self-employed in retailing. One way to do this was to lease and operate a service station. The number of self-employed dealers increased so rapidly in the thirties that despite a 50 percent reduction in secondary outlets, the amount of gasoline pumped per retail outlet fell by about 25 percent.[16]

As a result of World War II and the attendant price controls on, and rationing of, gasoline, the expansion of the retail market for gasoline and the increase in the number of stations was reversed. By the end of the decade, however, the retail market was once again rapidly expanding, about to enter its period of greatest growth.

14. A slight majority of all pumps, however, were located in service stations by 1927. See Hogarty, "Gasoline Marketing."

15. Hogarty, "Gasoline Marketing," p. 13.

16. Hogarty, "Gasoline Marketing," pp. 21–22. Hogarty quotes census figures showing that between 1929 and 1939 the number of retail establishments increased by 20 percent, and then fell by 3 percent between 1939 and 1954.

The Great Expansion 1950-1970

A number of new features marked the gasoline market in the fifties and sixties. One was the introduction of the super service station. Such a station had three or more islands, which allowed more rapid service and lower prices because of high volume. Some of the stations could also operate on a self-service basis.

A second innovation—which in a sense was merely the culmination of a trend that began early in the history of gasoline marketing—was the emergence of a number of nationally advertised brands of gasoline. The major oil companies that developed these brands stressed not only the quality of their gasoline but also the quality of the service provided at their stations and the extent of their dealer networks. Closely associated with the creation of nationwide dealer networks was the introduction by the major companies of credit cards that could be used to make purchases at their stations. Along with the nationally advertised brands, there was a parallel growth of local and regional chains of discount, high-volume, low-service retail outlets. Self-service was particularly common at such stations.

The enormous expansion of the service station network in the fifties and sixties resulted from the remarkable growth during those decades in the ownership and use of automobiles. New superhighways were built in the major metropolitan areas that led to a dispersal of population from central cities; an interstate highway system that encouraged long-distance driving for business and pleasure was largely completed. Moreover, the real price of gasoline was falling throughout the period. This decline reduced the cost of driving and increased the demand for fuel-intensive automobiles.

I have already maintained that the economic rationale for brand names in the retail gasoline market was necessary to create incentives for providing the quality of gasoline that consumers are willing to pay for. That rationale, however, does not explain why brand names became national in scope. The majors operate nationally, but they would not necessarily have to market their gasoline under national brand names. Nor does it explain why such brand names were not developed by nonintegrated marketing companies that would ensure the quality of their products just as department stores provide assurance of the quality of the products they sell but do not produce.

To understand these characteristics of the gasoline market, we must recall developments in the use of automobiles and in mass communications and advertising that were occurring almost simultaneously. As the amount of long-distance driving increased during the fifties and sixties, the average motorist became ever more likely to find himself driving in areas with which he was not very familiar. To such drivers, a familiar brand name upon which they could rely for quality products and especially quality service was particularly valuable. A merely local brand was not a good substitute to such drivers. Even the knowledge that a brand name for quality products and service was well established locally might not convey the information required by a motorist who lived elsewhere. If the service-station operator could identify the motorist as an outsider, he would have an increased incentive to provide him with poor products and service because such behavior would result in very little damage to the brand name of a local retailer.

Only if the brand were familiar over a substantial part of the country would the incentive of a local dealer to cheat outsiders be offset by the knowledge that to do so would depreciate the brand name and reduce the demand for the products sold under the brand name in other areas. A national company would thus have an incentive to police its dealers to ensure that such cheating did not occur. As the amount of long-distance driving increased, so did the demand for national, as opposed to merely local, brands of gasoline. Since the major oil companies were already operating on a national scale, it was profitable for them to expand their marketing operations—in many cases consolidating separate local brands into a single national brand.

The costs of establishing such national brands were reduced considerably by the advent of nationwide television networks, which afforded opportunities for national advertising. By so doing, the majors could familiarize the public with their brands and establish reputations for reliable products and service. "You can trust your car to the man who wears the star" was the slogan of the most successful gasoline marketer of that period. It was also one of the first well-known advertising slogans and conveys perfectly the message that Texaco and its major competitors were striving to deliver. National advertising did not merely spread a brand name from one area into another, it created a brand name signifying assured quality by virtue of the fact that it was advertised nationally. A company willing to invest in nationwide advertising is demonstrating its stake in the performance of its dealers and its willingness to risk the adverse consequences poor performance in one station would have on its brand name and sales elsewhere.[17]

The franchised dealer proved to be an ideal component in a national network of service stations. The franchisee was an independent business operator with a personal investment at stake in the station. Franchisees would have a greater incentive to increase sales than would employees. To ensure that such expansion did not come at the expense of reduced quality of service that would depreciate the value of its brand name, the franchisor would require a dealer to sell only its own products. Dealers would also be required to operate clean stations, provide routine service free of charge, and make other amenities available, such as snack machines and free road maps.

Although the franchisee would have some incentive to avoid providing the services the franchisor wanted provided (which incentive required the franchisor to incur costs to monitor the franchises), employees have similar incentives to avoid providing services and would have less incentive to increase sales. Employee-operated stations might thus have been more costly than dealer-operated stations for a company to monitor. Besides which, the franchisee's personal investment in the station and position as a local business owner provided the opportunity and the incentive to acquire knowledge of local conditions that would facilitate the adoption of effective marketing techniques and the posting of competitive prices. An employee would be less likely to have either the incentive or the opportunity to acquire or use such knowledge.

All this does not mean, however, that the dealer-operated station would always be superior to an employee-operated station. It does mean that the proportion of employee-operated stations in a company's network of dealers would depend on the relative costs and benefits to the company of the two types of stations. We shall see presently how recent alterations in the gasoline market have changed these costs and benefits, causing a changing mix of the two kinds of stations.

In return for providing service and products of an assured quality, retail outlets could sell nationally advertised brands of gasoline for pump prices that were somewhat higher than those of independent or local brands of gasoline. It was this premium that provided the incentive to incur the added costs of creating and maintaining a brand

17. See Klein and Leffler, "Market Forces in Performance."

name. But since not all motorists valued the additional services or the information about the quality of the products provided by the brand-name dealers, the market had room for the independent or local brands (some of which were actually owned by the majors) that provided less service to customers at a lower price than did major brand dealers. Thus at the same time networks of major brand-name stations were expanding, local brand, discount retail outlets operated by independent refiners, by independent marketers, and by some majors were also growing. Such stations were more frequently company owned and operated than were the major brand-name stations since they did not provide the kinds of services for which dealer operation was particularly suited.

The Post–1970 Contraction

During the seventies, particularly after dramatic increases in gasoline prices began to influence the driving habits of the public—the network of gasoline stations started to shrink. Those most severely affected were franchised dealers, but while the number of gasoline stations declined, self-service stations (usually employee operated) increased in number.[18]

That the number of self-service stations increased while dealer-operated stations were becoming less numerous was no coincidence. The two trends resulted from the same forces. First, self-service technology improved significantly, allowing a single attendant to supervise ten or more self-service pumps. In an environment of increasing costs for the largely unskilled labor employed at service stations, the savings made possible by self-service were substantial. Second, the demand for the services that service stations had routinely provided free of charge as well as for the minor repairs and maintenance they had offered declined during the seventies. This was caused by improvements in car design that increased the intervals between routine maintenance and eliminated the need for frequent battery, water, and tire checks and even the need for window cleaning. As motorists' demands for such services declined, they grew less willing to pay the premium for gasoline sold by full-service dealers. Another factor that made service stations less profitable was the increased competition

18. See Hogarty, "Gasoline Marketing," pp. 50-55.

they faced in the provision of minor service and maintenance from automobile dealers, automobile parts retailers, and even some department stores.[19]

The implications of these changes were even more far-reaching than just changing the proportion of self-service to full-service stations and employee-operated to dealer-operated stations. If the function of a brand name is to provide the customer with information about the quality of the product or service she can expect to obtain from a particular seller, then it is obvious that the brand name provides less information about a self-service dealer than about a full-service dealer. The brand name may still have some value even under self-service, but it surely cannot be as valuable to a customer at a self-service station as it would be at a full-service station. Thus stations with unfamiliar brand names have become increasingly good substitutes for stations with major brand names.

This suggests that the shares of the major brands in the retail gasoline market may have been, and are likely to continue, declining. The evidence confirms that this is just what has been happening. Between 1969 and 1980, the combined market share of the top eight gasoline marketers fell from 56.7 percent to 49.3 percent.[20] Following President Reagan's decontrol of the gasoline market in January 1981, which freed refiners from their obligations to continue supplying their customers during an earlier time period, the major oil companies began withdrawing rapidly from retail markets around the country.[21] By the third quarter of 1981, the share of the top eight firms had declined to only 46.7 percent of the retail market for gasoline.[22]

A national brand name will probably still continue to have considerable value to motorists who are far away from home – particularly when patronizing stations on the open road that do not rely on repeat business and from which motorists might require service or products in case of emergency. My conjecture is that in close proximity to major highways and resort areas the value of a nationally known name will continue to be such that stations carrying a major brand will retain a large market share. However, since long-distance

19. U.S. Department of Energy, *State of Competition*, pp. 191–205.

20. American Petroleum Institute, *U.S. Energy Markets*, p. 14.

21. See *The Wall Street Journal*, 12 February 1981, p. 2, and the *Oil and Gas Journal*, 5 March 1981.

22. *Lundberg Survey*, 23 December 1981.

driving is being displaced by air travel, that segment of the market has been shrinking and seems likely to continue to do so.

As the primary victims of these trends, franchised dealers have sought to cast the major oil companies in the role of villain. The dealers charge the majors with having conspired to drive their own "independent" franchised dealers from the market with a view to achieving a monopoly in the retail market. Although the discussion so far suggests that other factors are behind the decline of franchised dealers, and although the declining market shares of the majors make the charge that they have been monopolizing the retail market somewhat dubious on its face, the charges have still attracted much attention and have been the subject of Congressional hearings and an exhaustive Department of Energy study into the state of competition in the retail market for gasoline.[23] I shall address the charges of predatory behavior that have been leveled at the majors later in this chapter, but it will be helpful first to consider some reasons that would account for vertical integration between refining and crude oil production besides those I have already suggested for vertical integration between refining and marketing.

VERTICAL INTEGRATION IN THE OIL INDUSTRY

The huge vertically integrated companies controlling the flow of oil from the wellhead to the refinery and the flow of products from the refinery to the final consumer present a powerful, and to many a frightening, image. This view evokes visions of unlimited power and control, a capacity to manipulate supplies at will, and an immunity from, if not a dominance over, the market forces that would come into play if economic power were less concentrated and if the various transactions between successive stages of production took place between separate entities rather than within self-contained enterprises. Moreover, the ominous overtones of this image have been most skillfully emphasized by antagonists—both inside and outside the industry—of the major oil companies.[24]

23. U.S. Department of Energy, *State of Competition.*

24. For example, Allvine and Patterson, *Highway Robbery*, and Blair, *Control of Oil.*

My previous discussion has, I think, already cast doubt on the popular stereotype of the giant integrated oil company that engages in no market transactions until it sells the final product to powerless consumers. No oil company, we have seen, is entirely self-sufficient. All engage in some buying and selling in the upstream markets for crude and products. None is immune to market forces that would impinge on nonintegrated firms.

However misleading the popular stereotype may be, the question of why the major oil companies should be vertically integrated at all must still be considered. In the previous section I suggested some explanations for integration between the refining and the retailing stages. These explanations also imply that such integration is likely to diminish in the future. But what about integration between crude production and refining?

Let me now suggest an economic basis for such integration.[25] One of the problems that a purchaser who anticipates continuing to demand a particular product or resource must consider is the possibility that the supplier might suddenly seek to alter the terms on which it will continue supplying that product or resource. If the purchaser has any specialized equipment that requires a continuous flow of inputs from the supplier, the returns attributable to the equipment are vulnerable to an opportunistic threat by the supplier to withhold the inputs being provided. Ordinarily a purchaser is protected against such threats by the availability of alternative sources from which to obtain the required inputs on about the same terms as the purchaser had been obtaining them from the current supplier. But when the purchaser is in some way tied to the current supplier because of some specific capital investment that would make it very costly to switch to another supplier, the purchaser is vulnerable to exploitation.

Obviously a purchaser who foresees such a threat from a supplier would seek protection against potential exploitation. One possible alternative is to negotiate a long-term contract with the supplier so that the terms between them will be prescribed in advance. But long-term contracts have their own drawbacks. By prescribing the terms of future transactions in the present, they create potential conflicts should market conditions change unexpectedly. If the future price of the resource being supplied were to differ significantly from the price

25. See Klein, Crawford, and Alchian, "Vertical Integration."

prescribed in the contract, then the buyer or the seller would sustain a significant loss. If the contract is flexible enough to allow for alterations to reflect changes in market conditions or other unforeseen circumstances, it may not be sufficiently precise to prevent the exploitation the contract was intended to prevent.

A party who wishes to break a contract can always plead that circumstances have changed so that the terms of the contract should no longer apply. Moreover, long-term contracts must eventually be renegotiated. If the purchaser is tied to the supplier so that switching suppliers would be costly, the supplier may be able to exploit the purchaser when the contract comes up for renewal. This does not mean that long-term contracts are incapable of resolving the exploitation problem in all cases, only that this solution is not costless and that it entails certain risks of its own.

Under some circumstances, therefore, devices other than long-term contracts may be adopted to avoid the exploitation problem. One of these devices is vertical integration between the purchaser and the supplier. When the two are merged, the incentive for exploitation as well as the need to negotiate, and perhaps enforce, the provisions of a long-term contract vanish.

Vertical integration is not costless either, however. For example, it is certainly costly to merge separate firms. The incentives for integration may be so great, though, that both operations may be combined in a single firm from the outset. Then again, it is also possible that the two operations require different types of managerial skills or expertise so that it would be difficult to carry out both operations efficiently within a single firm.

Again the point to grasp is that the calculus of varied costs and benefits may dictate one solution in one particular case and another solution in another. Without detailed knowledge of the circumstances, one can do little more than indicate what some of the relevant considerations are in attempting to cope with the problem.

This analysis of opportunistic exploitation bears directly on the issue of vertical integration of production, transportation, and refining in the oil industry. A refiner of any substantial size normally has an enormous capital investment at stake in its refining equipment. An interruption in the flow of crude oil to a large refinery will quickly cause the entire operation to shut down, and the losses incurred from an unplanned shutdown are great. Thus any refiner tied to a specific pipeline delivering crude oil who does not have easy access to another source of crude has a strong incentive to own the pipeline. Any producer of crude oil tied to a specific pipeline and having no good alternative means of transporting its crude also has a powerful incentive to own the pipeline. This explains why joint ownership of pipelines by producers and refiners is so common in the oil industry. It likewise explains why refiners would seek to find and develop their own sources of crude oil.

A good test of the validity of this explanation of vertical integration and of its applicability to the oil industry is the extent of ownership by oil companies of oil tankers. Since no customer is in any way tied to a particular tanker, the opportunity for the owner of a tanker to exploit a customer is effectively limited by the availability of other tankers at comparable rates. Thus according to the explanation of vertical integration presented here, there would be little incentive for oil companies to own and operate their own oil tankers. Widespread ownership of oil tankers by oil companies would refute the hypothesis that vertical integration is a response to the exploitation problem.

The fact is that oil companies do not generally own their own oil tankers. Tankers are owned by separate shipping enterprises from whom the oil companies hire transport services. Thus the exploitation hypothesis has at least some empirical support.

PREDATORY BEHAVIOR IN THE GASOLINE MARKET

Charges of predatory tactics in the oil industry go back a long way, but they are still unproven. The earliest charges were those directed against John D. Rockefeller, who was accused of underpricing his competitors in order to drive them from the market or to force them to sell out to him at very low prices. Having eliminated his competition, it is said, he then was able to raise prices to monopolistic levels. Although the predatory tactics of the Standard Oil Company under Rockefeller have for years been part of the folklore of American capitalism, there doesn't seem to be any recorded evidence that Rockefeller actually used predatory tactics to monopolize the oil industry. What researchers have found is that Rockefeller paid very handsomely to buy out his competition. Indeed, he paid so well that some of his competitors seem to have reentered the market in order to sell out to Rockefeller a second time.[26]

The current charges against the major oil companies generally allege that they have been engaging in two types of predatory tactics. In either case the majors are supposedly using profits earned upstream to finance their predatory behavior downstream. One of the charges made is that the majors have used their upstream profits to subsidize their retail networks, in particular their franchised dealers, to drive independent marketers from the retail market.[27] This charge was quite common until the early to middle seventies in connection with gasoline price wars which were viewed as a predatory tactic used by the majors against competition.[28]

The other type of predation the majors have more recently been charged with is using employee-operated stations to drive out dealer-operated stations—the putative beneficiaries of earlier predation. The motive ascribed to the majors here is the creation of a monopoly in the retail market by eliminating dealers who are supposed to set prices more competitively than employees who set whatever price the oil companies prescribe.[29]

Before addressing the substance of these charges, I should like to make two preliminary observations. First, if the majors have been guilty of predation, they certainly have not yet been reaping the rewards they must have been expecting. Presumably their goal in selling below cost at retail was eventually to establish a retail monopoly that would allow them to sell above cost. But judging from the persistence of the charges of predatory competition directed at the majors, one would gather that they have been constantly incurring the costs of predatory competition at retail without having ever reaped the rewards.

The second point is that the entire argument rests on the assumption that the majors are in collusive agreement with one another that would allow them to enjoy monopoly profits if it weren't for the nonmajors and independents in the market who spoil everything by

26. See McGee, "Predatory Price Cuttings."

27. See Department of Energy (1981) for an exhaustive review of the evidence concerning charges that the majors have engaged in predatory competition.

28. Allvine and Patterson, *Highway Robbery*, Kardatzke, "Gasoline Price Fluctuations," and U.S. Department of Energy, *State of Competition*, pp. 5-6, 8.

29. U.S. Department of Energy, *State of Competition*.

acting competitively; hence, the majors try to act in concert to get rid of them. However, the assumption of ongoing collusion between the major oil companies is both implausible and inconsistent with evidence we have that bears on the extent of collusion in the oil industry.

The assumption is implausible because the number of major firms is so large and the market shares of even the largest of them are so small as to make effective collusive behavior in the absence of a formal mechanism to ensure compliance highly dubious. Moreover, the ineffectiveness of attempts to collude in the oil industry was clearly demonstrated in the late fifties when, under the voluntary import quota system, the major oil companies were encouraged by the government collusively to restrict their imports of crude oil into the United States and thereby to maintain higher prices domestically than in the rest of the world. Even with government approval and support, the agreement to restrain imports was ineffective and—in the absence of a formal government enforcement mechanism later provided by the mandatory quota system—quickly broke down.[30]

But even if, for the sake of argument, we grant that the majors collude effectively, it would still not be possible to suggest any plausible reason why they would engage in predatory competition. Consider first the allegation that the majors were seeking to establish a monopoly at retail by charging retail prices that did not cover their marketing costs. There are two possible cases—one in which we assume that the majors did not already have a monopoly at the refining stage, the other in which we assume that they did. Suppose they did not; then there simply is no end that could have been served by driving out competitors at retail. Unless sources of supply can be cut off to competing retailers (which is impossible without a monopoly in refining), the minimal costs of entry into the retail market would permit a steady flow of newcomers to replace the ones that were driven out just as soon as the majors raised their prices at retail above their marketing costs (presumably the object of the whole exercise).

So if the charge of predatory competition is to be raised at all, we have to assume that the majors already had a monopoly at the refining stage. In that case the argument must be that they were trying to extend a monopoly they had in refining to retail. Unlike the previous case, it would be possible to achieve a monopoly at retail by engag-

30. See discussion in Chapter 7.

ing in predatory competition in this one. But the mere possibility of extending a monopoly in refining to retail is not sufficient to make the predation charge credible. A motive must also be demonstrated, but none exists. Already possessing a monopoly at the refining stage, the majors would gain nothing from one at retail. Whatever monopoly return can be obtained from a market is gained once any single stage, whether upstream or downstream, has been monopolized.[31] Thus to suggest that the majors were engaging in predatory competition to extend a monopoly position held in the refining stage to the retail stage is tantamount to saying that the majors were giving money away to consumers (by charging them reduced prices) without any expectation of earning increased profits later on. You don't have to be an apologist for the oil companies to recognize how absurd that suggestion is.

Precisely the same argument disposes of the other charge of predation of which the majors have been accused. Either the majors have a monopoly in refining or they do not. If they do, they don't have to bother trying to extend it to retail. If they don't, trying to achieve a monopoly at the retail level is a futile and costly enterprise because entry into the retail market cannot be foreclosed.

CONCLUSION

Markets are far more complex and sophisticated mechanisms than textbook models would indicate. In the standard textbook model of COE, it is simply assumed that competition only affects the terms of trade—the price—at which exchanges are made. But contracts and exchanges in the real world are multidimensional. Contracts may involve a wide range of explicit and implicit mutual undertakings and obligations. Thus competition is rarely limited to competition in price alone; competition emerges within the entire range of contractual relationships between transactors so that when some contractual arrangements are found to serve the interests of the contracting parties more effectively than others, the more effective ones begin to achieve increasingly widespread acceptance.

Phenomena such as brand names, franchised dealers, vertical integration have evolved in a large number of markets in competition

31. See Bork, "Vertical Integration." The analysis has by now been incorporated into numerous textbooks. See for example Baird, *Prices and Markets*, pp. 250–53.

with other possible contractual forms because they have been found useful in coping with the practical problems of doing business. This is as true of the gasoline market as of others. To point out the fact that these and other sorts of business arrangements do not correspond neatly to the simple textbook model of COE is not an indictment of those practices. It is not necessarily an indictment of the model either, though it is an indictment of misuse of the model.

Thus charges that these practices are mechanisms for exploiting consumers, suppliers, or competitors usually cannot withstand critical scrutiny. But it is true that when certain business practices or arrangements have evolved under one set of conditions, a sudden change in circumstances may cause hardships to those committed to a particular way of doing business no longer adapted to current conditions. As we have seen, franchised dealers in the gasoline market have suffered from this kind of change in circumstances over the past decade or so. In the next chapter we shall see that other segments of the oil industry have also been suffering from various unforeseen changes.

There is always a strong incentive for those who lose from such unforeseen changes to seek compensation for their losses in the political process. This incentive seems to have been especially powerful in the oil industry. Unfortunately PC is both more wasteful and less adaptable than COE. So while easing the burden of adjustment for some, PC has made the overall adjustment to new circumstances in the gasoline market and the oil industry far more difficult than it need have been. The following chapters will show just how difficult.

Chapter 5

PRICE CONTROLS ON GASOLINE AND OTHER REFINED PRODUCTS BEFORE THE EMBARGO

> Whoever examines, with attention, the history of the dearths and famines which have afflicted any part of Europe, during the course of the present or that of the two preceding centuries . . . will find, I believe, that a dearth never has arisen from any combination among the inland dealers in corn, nor from any other cause but a real scarcity, occasioned sometimes . . . by the waste of war, but in by far the greatest number of cases, by the fault of the seasons; and that a famine has never arisen from any other cause but the violence of government attempting, by improper means, to remedy the inconvenience of a dearth.
>
> Adam Smith*

INTRODUCTION

Since 1973 Americans have lived in fear of an energy crisis that would cause a recurrence of the gasoline shortages they endured in the fall and winter of 1973–74. Thus for most of the ensuing decade two of the nation's paramount objectives were ostensibly to prevent the shortages from recurring or, if they did recur, to allocate supplies as efficiently and fairly as possible among competing users. The irony in this public concern about shortages was that a recurrence of the

*From *The Wealth of Nations*, pp. 492–93.

shortages was possible only because it was public policy to control the price of gasoline. Without controls shortages would not have been possible.

How is one to explain this apparent conflict between intentions and actions? I don't think it can be done without considering the role of PC in the creation of shortages. Controls on gasoline initially imposed for reasons that had nothing to do with the markets for petroleum products, disrupted COE for petroleum products and thus increased the incentive for organized groups to engage in PC for them. The resort to PC only led to the further breakdown of market allocation. With market forces hampered by government controls, the restriction of supplies caused by the Arab oil embargo inevitably led to shortages. The resulting chaos intensified PC for supplies.

In that case, why were controls maintained? Partly because by that time too many organized groups had acquired a stake in the controls owing to the benefits the controls and the associated mandatory allocations allowed them to extract from others. That many just did not understand the relationship between controls and shortages played a part too, but there were those who did understand and who clearly did benefit from a system of political allocation even though it produced shortages.

In this chapter I am going to recount how controls were imposed on the markets for gasoline and refined products and how, as a result, competition for these products became increasingly politicized. I shall cover the history until the imposition of the Arab oil embargo and the shortages that occurred in its wake. In Chapter 6 I shall explain how controls operated in the years after the embargo until they were finally lifted in 1981.

ORIGINS OF PRICE CONTROLS

In the summer of the third year of his first administration, Richard Nixon was, as we now know from various sources, hardly overconfident of reelection. Besides carrying the burden of an unpopular war that he had promised to end but had not, he had also presided over the first full-fledged recession in almost a decade. Yet despite the recession, inflation had fallen only modestly and was still running at what then was considered the rapid rate of 4 percent a year. Perhaps even more disturbing to Mr. Nixon was the subsequent weak recovery that had failed to bring down the rate of unemployment from its highest level since 1964.

Mr. Nixon had always blamed his loss to John Kennedy on a tight money policy that had prolonged the 1960 recession. He was therefore determined not to run for reelection encumbered by a recession caused by a tight money policy. But Mr. Nixon's plans to promote expansive fiscal and monetary policies prior to his reelection campaign were seriously threatened in the summer of 1971 by a looming crisis of confidence in the dollar.

Let me pause here briefly to explain how that crisis came about. In 1944 at a famous meeting at Bretton Woods, New Hampshire, the United States and its major allies established the International Monetary Fund and adopted a set of rules embodied in the fund's charter that were to govern international monetary relationships after World War II. Chaotic international monetary conditions during the interwar period had been widely held responsible for the Great Depression and the political upheavals of the thirties. So by creating a stable international monetary system, the architects of the Bretton Woods agreements hoped to help avoid a repetition of the bitter post–World War I experience.

The Bretton Woods system, as it came to be known, was anchored on a commitment by the United States to peg the price of gold at $35 per ounce. This commitment obligated the United States government to buy or sell gold at that price whenever foreigners wished to exchange gold for dollars or dollars for gold. Since United States citizens were legally prohibited from owning gold bullion, however, the dollar was not fully convertible into gold as it had been under the pre–World War I gold standard.[1] Other countries fixed exchange rates between their own currencies and the dollar. Although countries were obligated to maintain these parities, they were permitted to alter them in the face of persistent balance of payment deficits or surpluses. For about twenty years after the war, the Bretton Woods system worked reasonably well. The United States, which produced the dominant international currency, maintained reasonable price stability over the period as it was supposed to. The persistent U.S. balance of payments deficit and a modest decline in U.S. gold holdings signified neither misbehavior by the U.S. monetary authorities nor a threat to continued convertibility. Rather they were necessary

1. See Friedman, "Gold Standards."

consequences of an expanding world economy in which there was growing foreign demand to hold both dollars and gold. Thus, the U.S. exported the dollars and the gold demanded by the rest of the world and imported goods and services with which the rest of the world paid for the dollars and gold it was acquiring.

In the middle and late sixties, though, expenditures for the Vietnam War and the Great Society programs began to exceed tax revenues by substantial amounts. These deficits were financed by an increasing rate of monetary expansion in the United States. The excess supply of dollars began to drive up domestic prices, but these dollars were also exported abroad to pay for increased imports and foreign investments. Unlike the dollars that had been exported earlier, which foreigners had willingly held, the dollars now being exported were in excess of the demand by foreigners to add to their dollar balances. So foreigners sought to redeem excess dollars for their domestic currencies or for gold. By 1968 the demand by foreigners to redeem dollars for gold led to a run on U.S. gold reserves. The United States then stopped selling gold to anyone except foreign central banks. Even these banks were told quietly that the United States was no more disposed to sell gold to them than to anyone else. Unable to cash in all their excess dollars for gold, foreigners instead sought to cash in the dollars for their domestic currencies at the fixed exchange rates maintained by their governments. Thus foreign governments and central banks began accumulating huge reserves of unwanted dollars. By 1971 it was becoming clear that the situation had gotten out of hand.[2]

Perhaps spurred by expectations that Mr. Nixon would seek to stimulate the U.S. economy by loosening monetary policy, the free market price of gold began rising, and the drain on U.S. gold reserves became increasingly rapid. Faced with an incipient international monetary crisis, Mr. Nixon had three alternatives. First, he could maintain (or, better, restore) convertibility at $35 per ounce. But to do so would have required such drastic monetary restraint that a severe recession would have been inevitable and his hopes for reelection would be dashed. A second alternative was to restore convertibility at a higher gold price so the incentive to redeem dollars would be eliminated. Although this option would have avoided a recession,

2. See H. G. Johnson, "Monetary Reform," reprinted in H. G. Johnson, *Economics and Society*, pp. 248-66.

it would have compelled Mr. Nixon to take formal responsibility for devaluing the dollar, which would have been politically embarrassing for him at home and abroad. Mr. Nixon's third alternative was to abandon convertibility altogether. To do so would halt the drain on U.S. gold reserves without compelling him to cause another recession or to take responsibility for a formal devaluation of the dollar. Instead, he could announce a decision aimed, as he put it, at preventing international currency speculators from holding the dollar hostage.

THE WAGE-PRICE FREEZE AND PHASE II

The bold stroke, the radical departure, and the daring gamble were always appealing to Mr. Nixon. In July 1971 he stunned the world by announcing his visit to China. Thus it was not completely out of character for him to have decided, only a few weeks later, to abandon the vestiges of the dollar's convertibility into gold. In so doing, he freed himself from the international constraints on his plans to pursue expansionary fiscal and monetary policies prior to and during his reelection campaign.

Nevertheless Mr. Nixon still faced a problem of domestic inflation that had proved less tractable than he and his advisors had hoped or anticipated. They also realized that an expansionary monetary policy would reignite inflationary pressures that, after imposing considerable hardship, had only begun to subside. If the abandonment of convertibility led to expectations of more rapid inflation, expansive monetary and fiscal policies would not lead to the increased output and employment Mr. Nixon wanted. Thus, hoping to reduce expectations of inflation, Mr. Nixon and his advisors decided on another, perhaps bolder, stroke—a ninety-day freeze on all wages, prices, rents, dividends, and interest rates. During this time a program for continuing wage-and-price controls would be developed. Mr. Nixon announced the freeze and the abandonment of convertibility on August 15, 1971.

At first, petroleum prices were subject to the same regulations as other prices (though later special regulations would be adopted for petroleum). Nor, to begin with at least, was the impact of the freeze and the phase II controls on petroleum markets very disruptive. In 1971 domestic crude oil prices were still comfortably above the world price so there was, as yet, little upward pressure on domestic crude prices. The price of gasoline, at its seasonal peak near the end of the summer driving season when the freeze was imposed, was about to begin falling. Heating oil prices, however, were at seasonal lows and normally would have been about to start rising. Luckily, heating oil stocks were unusually high, and the fall and winter of 1971–72 were rather mild so that demand for heating oil could be met even without the added production that normally is induced by the seasonal increase in heating oil prices.

With the end of the ninety-day freeze, phase II of the controls began on November 14, 1971. Price increases were permitted only when increased costs had been incurred. Profit margins were not permitted to exceed the average profit margin of the best two of the previous three years regardless of whether a firm's prices had increased.[3] This meant that once their profit-margin limits had been reached, firms no longer had any incentive to reduce costs.

Even though the phase II controls did not specifically discriminate against the oil industry, the Price Commission, which was supposed to interpret and enforce the phase II regulations, denied the oil industry the benefit of a provision of the controls to which the industry was arguably entitled. The regulations required firms to compute their ceiling prices by using their prices on August 15 as the base upon which to add cost increases incurred after November 14. But in cases in which price fluctuated seasonally, it was permissible to calculate base-period prices using the price on some date prior to August 15. Despite the seasonal pattern of gasoline and heating oil prices, oil companies were not permitted to alter their base-period prices accordingly.[4]

Because of the large inventories of heating oil and the mild weather, the adverse effects of not permitting a change in the base-period price of heating oil were delayed until the following winter—after the 1972 elections. Throughout 1972 heating oil prices were abnormally low relative to gasoline prices. As a result refiners increased gasoline yields at the expense of distillate. By summer concern about the heating oil supplies for the winter began to grow. Government officials exhorted the oil industry to increase heating oil output—notwithstanding the admitted disincentive of a low ceiling price for

3. Owens, "Petroleum Price Controls," p. 1236.

4. W. A. Johnson, "Impact of Price Controls," p. 102; Allvine and Patterson, *Highway Robbery*, p. 140–48.

heating oil. The oil industry, it was said, could afford the sacrifice necessary to ensure that supplies of heating oil were adequate to forestall shortages.

These pleas fell largely on deaf ears, which may not say much for the public spiritedness of the oil companies, but it does testify to their competitiveness. If the oil companies do compete with each other, then no firm is likely to sacrifice its own profit to confer a benefit on its competitors. But that is just what the oil companies were asked to do. Producing more heating oil was unprofitable for each individual firm, but it would have benefited the entire industry by improving the industry's public image and its reputation for good citizenship and corporate responsibility. But no individual company, by itself, could have increased heating oil output enough to have made much difference in the total supply of heating oil. Thus it would have only been through some cooperative agreement characteristic of an industry cartel that such an increase in heating oil output could have occurred.

PHASE III AND SPECIAL RULE NUMBER ONE

Phase II came to an end on January 10, 1973. It had succeeded admirably in preventing the inflationary pressures created by Mr. Nixon's campaign to reelect himself from becoming manifest prior to the election. The consumer price index rose by only 3.4 percent in 1972. However, the controls that had been repressing those pressures were also beginning to cause shortages of a wide variety of products and resources. To prevent these shortages from disrupting economic activity, it was necessary to relax or eliminate the controls, but doing this meant letting loose repressed inflationary pressures.

Phase III was supposed to steer a course between Scylla and Charybdis. To avoid shortages, controls under phase III were relaxed in three respects. First, firms were allowed to increase their profit margins to the average of the best two out of the previous five (instead of previous three) years. Second, firms that increased the weighted average of their prices by less than 1.5 percent annually were freed from all profit-margin controls. Third, there was no requirement for firms to obtain permission for price increases provided those increases were within Phase III guidelines.[5] Nor did firms have

5. W. A. Johnson, "Impact of Price Controls," pp. 103–04.

to submit documentation to CLC to justify price increases. There was merely a requirement that records be kept in such a manner that the justification for any price increase could be checked by CLC. The administration called phase III "voluntary and self-administering," using the income tax as a model. Without an effective enforcement mechanism, however, the only real deterrent to noncompliance was the possibility that CLC might hold hearings to investigate a price increase and thus cause the firm adverse publicity. If it were found to have violated phase III guidelines, the firm would probably not be required to do more than rescind the price increase and refund overcharges.

Despite the relaxation of controls, it was hoped that phase III would somehow continue to restrain rising prices. But once phase II controls were lifted, heating oil prices, in the face of incipient shortages, rose rapidly from the levels at which they had been virtually frozen since 1971. The increases in heating oil prices could be justified under phase III because refiners had already absorbed higher costs stemming from rising world crude oil prices. Nevertheless, the rapid increases in heating oil prices led to bitter protests from consumers and politicians, particularly those from the Northeast where the price of heating oil had long been a major political issue. The outcry against higher prices compelled CLC to hold public hearings to investigate the price increases. No finding of any violation of phase III regulations was made, but to placate the political opposition to higher heating oil prices, CLC issued special rule number one (SRNO).[6]

Only the twenty-three largest firms in the oil industry, those with annual sales over $250 million, were covered by SRNO. The other tens of thousands, if not hundreds of thousands, of firms involved in one stage or another of the oil industry were exempt from SRNO and, in effect, decontrolled under phase III. Firms covered by SRNO were permitted to increase the weighted average of the prices of products and crude oil sold to third parties by up to 1 percent over the weighted average of their prices on January 11, 1973. Increases between 1 and 1.5 percent had to be supported by evidence of cost increases incurred after March 6, 1973 (the date SRNO went into effect). Increases over 1.5 percent were subject to profit-margin limitations and had to receive prior approval from CLC.

6. W. A. Johnson, "Impact of Price Controls," p. 104; Owens, "Petroleum Price Controls," pp. 1238–42.

Since prices had increased substantially after January 11, the firms subject to SRNO had already exceeded the 1.5 percent limit. World crude oil prices had also been rising steadily, and most of the companies covered by SRNO were producing substantial amounts of crude oil outside the United States. Consequently, they were exceeding their profit-margin limits, and companies had no choice but to roll back their domestic prices.[7] Despite market pressures to increase prices, firms subject to SRNO that produced significant amounts of crude oil outside the United States were compelled to reduce their domestic product prices. Since they could obtain higher prices abroad for refined products than they could domestically, such firms began to reduce the amount of crude oil they shipped to the United States.

The reduced imports of foreign oil by the larger companies had severe consequences not only for the covered firms but also for the uncovered firms. To understand why, we must consider how oil had been supplied to refineries under the mandatory oil import quota system. Under this system, which I shall describe in more detail in Chapter 7, rights to import crude oil (import-quota tickets) were distributed to refiners according to a formula (the sliding scale) that awarded proportionately more tickets to small refiners than to large ones.[8]

As long as the price of imported oil was significantly below the price of domestic crude, import-quota tickets were valuable assets. Every recipient of quota tickets was thus subsidized by the import-quota program; and because of the sliding scale, small refiners, usually located in inland areas, were proportionately the most heavily subsidized.

Inland refiners usually found it cheaper and more profitable to exchange their quota tickets for oil from a nearby source owned by integrated producer-refiners that operated a refinery near a coastal port. An integrated firm could save transportation costs by exchanging its inland oil for oil imported by the inland refiners. The inland refiner, in effect, obtained domestic oil at the cheaper world price. This savings was a subsidy to the recipient of the quota tickets. With-

7. W. A. Johnson, "Impact of Price Controls," p. 104; Lane, *Mandatory Petroleum Regulations*, pp. 14-17.

8. On the oil import quota generally see Bohi and Russell, *Limiting Oil Imports*; Dam, "Import Quotas"; and Barzel and Hall, *Oil Import Quota*. On the sliding scale in particular, see Bohi and Russell, *Limiting Oil Imports*, pp. 114-25; Dam, "Import Quotas," pp. 20-24; and Barzel and Hall, *Oil Import Quota*, pp. 51-64.

out this subsidy, many small inland refiners would have had to shut down.[9]

In early 1973 the rising world price of crude oil was wiping out the subsidy implicit in the quota tickets. Not only that, but SRNO was making it unprofitable for most of the larger companies to import crude oil into the United States. If they did import oil, it was only to use along with their own domestic supplies to avoid the costs of shutting down their domestic refining and marketing operations. Thus the majors were unwilling to continue supplying the small inland refiners with crude oil from domestic sources that the majors needed now to replace former imports rendered unprofitable by SRNO. Not only were they deprived of their subsidy, the small inland refiners were not even able to obtain as much unsubsidized crude oil as they demanded.[10]

In an attempt to reduce barriers to the importation of crude oil, President Nixon announced on April 18, 1973, the termination of the mandatory oil import quota program. But this step had little effect since SRNO had replaced the import quota system as the operative constraint on oil imports.[11] Nor could domestic crude production expand to make up for the shortfall in imports. Domestic production had peaked in 1970, and the spare capacity in Texas and Louisiana that had been maintained by demand prorationing was already exhausted.

SRNO was also causing havoc in product markets. Independent marketers had either produced products in their own refineries or purchased products from independent refiners or major refiners. Because of interruptions in the supply of crude, independent marketers were, in many cases, either unable to produce their own products or unable to purchase them from independent refiners. Moreover, because their prices were held down by SRNO, the major refiners found it profitable to transfer supplies from their relatively low-priced customers (usually independent marketers or other bulk purchasers) to their own jobbers and retail outlets to whom they charged higher prices. Jobbers and retail outlets previously would have been unwilling to purchase increased quantities from the majors, but that was now changed. Since jobbers and retail outlets were not

9. Bohi and Russell, *Limiting Oil Imports*, p. 77.

10. W. A. Johnson, "Impact of Price Controls," pp. 105–06; Lane, *Mandatory Petroleum Regulations*, pp. 16–17; U.S. Department of Energy (1981), p. 215.

11. Bohi and Russell, *Limiting Oil Imports*, pp. 218–19.

covered by SRNO, they were free to increase prices as much as market conditions allowed. What SRNO did do was to hold down the prices of their supplies. With rapidly increasing profit margins, they naturally sought to increase their sales and demanded as much as possible of the now low-priced products being supplied by the majors.[12] Indeed, in some cases jobbers and retailers went to court to invoke contractual obligations to compel their suppliers to provide them with as much output as they desired at the controlled prices.[13]

Suffering from sharply curtailed supplies of crude from the majors, nonmajor refiners rapidly bid up the price of the remaining crude still available on the market. Having also lost their subsidy from the import quota system, nonmajor refiners (especially the least efficient ones that had only survived because of subsidies from the import quota system) sharply raised the prices of their products. For independent marketers, traditionally the discounters in the retail market, this meant they often had to sell at prices much higher than the retail outlets of the major oil competitors.[14]

The independent refiners and marketers attributed their plight to a predatory conspiracy on the part of the major oil companies. Other critics of the major oil companies reached a similar conclusion.[15] Demands for government intervention to prevent the majors from destroying their competitors were made by the independents to Congress, and Congress quickly began working on legislation to force the major oil companies to share their supplies equitably with all their customers. The daring thought that the hardships inflicted on the independents had been caused by previous government intervention seems to have crossed few minds and passed even fewer lips.

One instance of how the disruptions in petroleum markets were attributed to predatory tactics by the majors in disregard of the role of government intervention is a study by Allvine and Patterson.[16] They maintained that the major oil companies had cut off supplies from independent refiners and marketers as part of a conspiracy against their competitors. To support their contention, Allvine and

12. U.S. Department of Energy, *State of Competition*, pp. 214–15; Lane, *Mandatory Petroleum Regulations*, p. 24.

13. For example, *North Penn Oil & Tire Co.* v. *Phillips Petroleum Co.*, 358 F. Supp. 908 (E.D. Pa. 1973).

14. W. A. Johnson, "Impact of Price Controls," p. 107; Bohi and Russell, *Limiting Oil Imports*, pp. 219–20.

15. For example, Allvine and Patterson, *Highway Robbery*, and Blair, *Control of Oil*.

16. Allvine and Patterson, *Highway Robbery*, p. 189.

Patterson adduced the drastic decrease in the amount of crude oil supplied to independent refiners by the majors in early 1973. However, they completely ignored the existence of SRNO and were thus oblivious to the significance of the fact that the most substantial cutoffs began to occur in March 1973 just as SRNO went into effect.

In May Congress passed legislation authorizing the President to impose mandatory allocations on crude oil and products. The administration, however, chose instead to implement a voluntary program that did not require the participation of the companies.

The situation under SRNO was aptly described by one study as follows: "Conditions had been created wherein there was no market incentive to raise domestic production, no incentive to reduce consumption, and no incentive to increase imports."[17] One need hardly add that under these circumstances shortages were inevitable.

THE SECOND FREEZE AND ITS AFTERMATH

Under phase III, the inflationary pressures fueled by the fiscal and monetary policies that had been pursued since the imposition of wage-and-price controls were more or less free to exert themselves. As a result, the annual rate of increase in the consumer price index during the first half of 1973 was about 8 percent, and the annual rate of increase in wholesale prices was over 20 percent.[18] Already deeply immersed in Watergate, then rapidly unfolding, Mr. Nixon again sought relief, if not escape, from a difficult economic and political situation by imposing another freeze on wages and prices on June 13, 1973. The second freeze was to last for sixty days during which time revisions of the phase III price controls were to be prepared.

In contrast to the first freeze, imposed when inflation was falling and the economy was sufficiently slack to keep the shortages and distortions that inevitably follow controls from becoming immediately apparent, the second freeze was imposed during rapid and accelerating inflation when no slack was left in the economy. Thus shortages and other distortions became evident almost immediately after the freeze was declared.

17. Bohi and Russell, *Limiting Oil Imports*, p. 218.
18. Brock and Winsby, "Removing Controls," p. 867.

One such distortion arose because the freeze allowed for the increased costs of imported petroleum products to be passed on to consumers only if the products underwent no physical change before being sold to consumers. Imported products had to be segregated from domestic products that were subject to the price freeze, and oil companies were required to sell domestic products at lower prices than identical imported products.[19] Resentment against the widely different prices that were arbitrarily charged various buyers (a violation of the Robinson-Patman Act's prohibition against price discrimination) was directed against the oil companies, not against the regulations. Proposals for mandatory allocations to ensure fair treatment of all customers gained further support.[20]

Even as the demand for tighter control over petroleum markets intensified, the effects of the second freeze were leading to widespread disillusionment with the whole idea of wage-and-price controls. Presidential authority to control wages and prices, which few now wanted to see continued, was scheduled to expire on April 30, 1974. Thus phase IV of the controls, which was to follow the freeze, was envisioned as basically a transition to general decontrol. To ease this transition, CLC was supposed to devise special guidelines for each sector of the economy that would somehow help to create the proper conditions for ultimate decontrol.[21]

The administration feared that sudden decontrol would be disruptive and cause excessive price increases in some sectors. Therefore, industries that appeared to the administration to pose inflationary threats, of which the oil industry was one, were subjected to tight controls under phase IV in a final effort to beat down inflationary pressures before the controls expired.[22]

The prices of crude oil and refined products had been rising rapidly all during 1973. In the second quarter gasoline and motor oil

19. W. A. Johnson, "Impact of Price Controls," p. 108.

20. Bohi and Russell, *Limiting Oil Imports*, pp. 220-21.

21. Brock and Winsby, "Removing Controls," pp. 868-71.

22. If the controllers were worried that sudden decontrol was going to result in massive price increases, then tightening controls in sectors in which upward pressure on prices was stronger was hardly likely to reduce the shock of ultimate decontrol. The only sense one can make out of the selective decontrol approach is that it was a way to spread out over six months price increases that would otherwise have occurred all at once. But if so, the controls should have been relaxed in the sectors experiencing the most rapid price increases. Tightening these only increased the distortions and the size of the ultimate adjustment after decontrol.

prices rose at a 26 percent annual rate as measured in the consumer price index.[23] Shortages of gasoline, heating oil, and propane had already become noticeable, though still not too serious, except for propane.[24] But the prospect of renewed heating oil shortages during the coming winter was already sending shivers down the spines of administration officials.

While the freeze was still in effect, CLC began to prepare the phase IV regulations for each industry that were to go into effect once the freeze expired. The regulations for the oil industry evoked immediate protests and legal challenges, primarily from retailers. As a result, the freeze on crude oil and refined product prices was extended until September 7, when the phase IV regulations for the oil industry finally went into effect.

The regulations allowed refiners to charge their prices as of May 15, 1973, plus any subsequent increases in their average costs attributable to increased prices for raw materials and products.[25] Wholesalers and retailers were permitted to charge their purchase prices plus either their absolute dollar markup as of January 10, 1973, or 7¢ per gallon, whichever was greater.[26] In order not to discourage imports of crude oil and products, refiners were permitted to commingle imported and domestic supplies for the first time and to count the prorated share of the more expensive imported supplies in calculating their ceiling prices.[27]

Owing to the administrative difficulties of applying two different standards to the retail and refining operations of integrated refiner-marketers, the retail outlets of refiners were subject to ceilings based on their May 15 prices and increased costs incurred after May 15. The difference in treatment of the integrated refiner-marketers and other retailers, whose ceiling prices were based on their markups as of January 10, was particularly resented by independent retailers. They viewed the January 10 markups as unfair and burdensome because price wars, resulting in abnormally low markups for many of

23. Owens, "Petroleum Price Controls," p. 1250.

24. See W. A. Johnson, "Impact of Price Controls," pp. 104-05, for a vivid description of how controls caused the severe shortage of propane in 1973. The propane shortage was one of the first product shortages to gain widespread national attention.

25. Owens, "Petroleum Price Controls," pp. 1266-68.

26. Owens, "Petroleum Price Controls," pp. 1269-70.

27. W. A. Johnson, "Impact of Price Controls," p. 108.

them,[28] had been widespread on that date. Nor could they recover any increases in their operating costs incurred after January 10.[29]

Retailers therefore launched an urgent campaign against the phase IV regulations, lobbying Congress and the administration for relief and conducting public protests and demonstrations while challenging the legality of the regulations in the courts. Although the U.S. District Court for the District of Columbia did rule in favor of the retailers in one lawsuit, the ruling was overturned on appeal. Other lawsuits were also unsuccessful, but continuing protests and demonstrations (including a nationwide shutdown by dealers) eventually forced Congress and the administration to order CLC to revise its regulations affecting the retail sector. The Findley Amendment, enacted on September 25, 1973, prohibited CLC from establishing a different base period for retailers than that established for refiners. The Findley Amendment also required that CLC permit full pass through of all crude petroleum, refining, transportation, and marketing costs actually incurred at any stage of the supply or distribution of refined products.[30] CLC attempted to limit the effect of the Findlay Amendment by ruling that such increased costs could only be passed through in 1¢-per-gallon increments. This would have required refiners, distributors, and retailers to absorb part of the increased costs they had actually incurred. An increase in average costs of say 1.5¢ per gallon would only permit an increase of 1¢ per gallon in a firm's ceiling price. However, this regulation also provoked the protests of gasoline retailers, and CLC had to relent on this point as well. The final attempt by CLC to limit the extent of pass-throughs was to allow changes in ceiling prices only once a month.[31] As we shall presently see, the once-a-month rule would play a crucial role in precipitating the shortages.

THE ARAB OIL EMBARGO

On October 6, 1973, Israel was attacked by both Egypt and Syria. To support Egypt and Syria, the Arab oil producing countries an-

28. Owens, "Petroleum Price Controls," pp. 1256-57.
29. Owens, "Petroleum Price Controls," pp. 1276-77.
30. Owens, "Petroleum Price Controls," pp. 1284-85.
31. Owens, "Petroleum Price Controls," pp. 1287-88.

nounced shortly thereafter that they would embargo oil shipments to the United States and the Netherlands so long as these countries continued to follow unacceptably pro-Israeli policies. In conjunction with the embargo, they also announced an immediate reduction of 5 percent in their daily output of crude oil. Their announcement caused a panic in world oil markets that sent the price skyrocketing. Within weeks the price of oil in spot markets climbed to over $10 per barrel; the oil producing countries of course soon began to increase prices on oil sold under long-term contracts correspondingly.

It was not long before shortages of gasoline became widespread in many parts of the United States. And because the United States had been singled out as the primary object of the oil weapon that the Arabs had decided to unleash, it was natural to assume that the shortages were the consequence of the embargo. But the embargo as such was irrelevant to the shortages. Oil is fungible, and a refusal by some to sell oil to the United States would simply mean a rearrangement, at some additional cost, of buyers and sellers. The Arabs certainly had no way of preventing non-Arab producers from selling to the United States or even preventing their own customers from reselling oil to the United States. So long as buyers in the United States were willing to pay as much for oil as buyers elsewhere were paying, the Arab oil embargo could have no effect on the United States. By reducing their total output, the Arab oil producers found that they were able to force up oil prices generally, but that in no way prevented the United States from obtaining as much oil as it was willing to pay for at prevailing prices. If you doubt this, consider the ineffectiveness of the attempt by the United States to embargo grain shipments to the Soviet Union after the Soviets invaded Afghanistan.

Shortages did not arise because the United States could not import crude oil. By the time of the embargo, refiners could pass through the costs of imported crude oil so there was no serious disincentive to importing crude oil. The main problem was the lag in passing those costs through to consumers—which was considerably exacerbated by the once-a-month rule. Facing rapidly rising costs and an excess demand for gasoline, dealers had a powerful incentive to sell supplies early in the month and to build up stocks later in the month in anticipation of an imminent price increase. This made it increasingly difficult for motorists to obtain gasoline as the end of the month approached.

Once the public began to anticipate that gasoline would not readily be available, a panic-buying syndrome quickly became widespread.

In an unconstrained market, rapidly rising prices would deter such panic buying, and the observation that supplies in fact remained available would dispel the expectations of shortages. Once the panic subsided, prices would decline. But under controls, panic buying resulted in long gasoline lines at the dwindling number of open gasoline stations as the month dragged on. Instead of being dispelled, as they would be in a free market, fears were only too quickly confirmed as shortages spread from one urban area to another in the late fall and winter of 1973–74. Merely the amount of gasoline consumed by motorists idling their engines while waiting in queues at gasoline stations or while driving in search of an open station was a substantial drain on gasoline supplies.[32]

Evidence that the once-a-month rule played a major role in creating gasoline shortages is provided by the pattern of fluctuations in gasoline inventories. From October to April there was a cyclical pattern in the fluctuations. Inventories fell early in the month and were built up later in the month. Under normal conditions, however, such fluctuations in inventory levels should have occurred randomly.[33]

Even before the embargo, there had been pressure for mandatory allocation of crude oil and refined products. Once the embargo was declared, the passage of a bill requiring the imposition of mandatory allocations became a foregone conclusion. Groups that had already experienced supply difficulties sought the protection of mandatory allocations to ensure their continuing access to supplies of crude oil and products. To establish mandatory allocation, Congress passed the Emergency Petroleum Allocation Act (EPAA) on November 27. The President set up the Federal Energy Office (FEO) to take the administration of the phase IV controls over the oil industry from CLC and to prepare, implement, and enforce the regulations and allocations required by EPAA. On January 15, 1974, the initial set of mandatory allocation regulations was published. The regulations froze existing relations in the distribution chain for crude oil and refined products. Producers were required to continue selling to refiners with whom they had been dealing on December 1, 1973, and refiners were required to continue supplying customers the amounts they had supplied during the corresponding month of 1972, the base period

32. U.S. Department of Justice, *Gasoline Shortage*, p. 162. The Justice Department study estimated that 144,000 barrels of gasoline per day were wasted by motorists idling in queues at gasoline stations during the 1979 shortages.

33. W. A. Johnson, "Impact of Price Controls," pp. 113–15.

of the allocation program.[34] Authority to maintain controls over petroleum prices (which was scheduled to expire along with authority to control wages and prices generally) was extended until August 31, 1975.

Besides requiring crude producers to continue supplying their customers as of December 1, FEO initiated the crude oil buy-sell program. Under this program refiners operating at a higher percentage of capacity than the national average were required to sell some of their crude oil to refiners operating at less than the average percentage of capacity. Those obligated to sell supplies under the program had to do so at their own weighted-average acquisition cost plus a fee. During the first three months of the program, 56 million barrels were exchanged, an average of over 600,000 barrels per day.[35]

The idea behind the program was to compel the majors, who were suspected of having withheld supplies from other refiners in the first place, to share their supplies with other refiners. However, not all majors were sellers under the program, nor were all independents buyers under the program. It turned out that some majors were short of supplies, while some independents were amply stocked.

Thus the program awarded certain refiners the right to acquire oil belonging to other refiners at below market prices. (Exchanges took place at less than market prices because the average acquisition cost included the controlled prices of domestic crude oil, while the market price corresponded to the price of imported oil. The average of domestic and imported prices was obviously less than the market price.) The less oil obtained on the market, the more oil a refiner was entitled to obtain at a discount. Similarly the extent of a refiner's obligation to sell oil under the program increased along with its purchase of oil on the open market. The perversity of these incentives is manifest. If a refiner were operating at a percentage of capacity above the national average, part of any additional crude oil obtained would have to be sold for less than the oil was worth.[36] If a refiner were operating at a percentage of capacity less than the national average, any additional crude oil obtained would reduce the amount of oil the refiner was entitled to purchase at a discount. Thus the buy-

34. Lane, *Mandatory Petroleum Regulations*, pp. 39–40.

35. Lane, *Mandatory Petroleum Regulations*, pp. 40–43.

36. Some of the distortions created by the buy-sell program were later removed. See Lane, *Mandatory Petroleum Regulations*, pp. 42–43.

sell program reduced the amount of crude oil refiners were willing to import, and as a consequence refiners produced less gasoline and refined products.

Fearing a recurrence in shortages of heating oil, CLC also encouraged refiners to expand the production of heating oil at the expense of gasoline. Thus, CLC permitted refiners to increase distillate prices if they increased distillate yields. Because of the "tilt" in their favor, heating oil prices rose by 49.3 percent between September 1973 and June 1974, while gasoline prices increased by only 32.7 percent.[37] The diminished consumption and increased output of distillate induced by rapidly rising distillate prices prevented the occurrence of any shortages of heating oil during the entire embargo period.[38]

Concern about shortages of heating oil also caused FEO to require refiners to build up excessive inventories of heating oil. Ignoring the effect that higher prices would have on the consumption of heating oil, FEO insisted that refiners accumulate inventories sufficient to meet a level of heating oil demand significantly greater than that of 1972. Distillate stocks at the end of February 1974 were 36 million barrels greater than they were at the end of February 1973.[39]

In fact the FEO ordered a general buildup in inventories of crude oil and other products. In September 1973 total U.S. petroleum stocks were 28 million barrels less than they had been one year earlier. By April 1974, when the increased costs arising from higher oil prices had been largely incorporated into the U.S. price structure, total petroleum stocks exceeded their levels of a year earlier by 52 million barrels.[40] Gasoline stocks alone were 20 million barrels greater in April 1974 than they had been in April 1973. In sum, FEO was responsible for the hoarding of available gasoline supplies and for the excess production of other products at the expense of gasoline.[41]

Having seen to it that not enough gasoline would be produced and following its Congressional mandate, FEO allocated whatever gasoline had been produced. In doing so, of course, FEO exacerbated the shortages. The allocation regulations conferred priority—or perhaps

37. Owens, "Petroleum Price Controls," pp. 1315-17.

38. Lane, *Mandatory Petroleum Regulations*, pp. 36-37.

39. Lane, *Mandatory Petroleum Regulations*, pp. 36-37.

40. Mancke, *Squeaking By*, p. 24.

41. Lane (*Mandatory Petroleum Regulations*, p. 47) estimates that gasoline production might have been increased by up to 237,000 barrels per day if heating oil stocks had not been increased over the previous year's level.

more accurately privileged—status upon certain groups of users that could compete effectively in the political arena. These included farmers, energy producers, providers of emergency services, sanitation services, telecommunications services, and transportation services. All of these were entitled to 100 percent of what was euphemistically called their "requirements." All other wholesale and end-use bulk purchasers were entitled to 100 percent of their base-period purchases. State governments were also allowed to create and reward favored categories of users, and each state government was authorized to allocate 3 percent of total supplies in the state to whichever users it judged most deserving.[42] Retail gasoline outlets received what was left over after all the priority users had secured their shares. No priorities were established to determine how much different retail buyers would obtain, and market prices were not permitted to ration the limited supply at retail. So the predominant mode of competition for the supplies that had not already been allocated through the political process became willingness to wait in queues outside gasoline stations.

Because allocations to nonpriority users were based on their purchases during the base period (1972), those parts of the country that had experienced the most rapid growth since 1972 suffered greater shortages than areas of the country that had grown less rapidly. Furthermore, once shortages became widespread, the expectations that they would continue induced new changes in demand that made allocations based on use in 1972 even more outdated. For example, to avoid being cut off from supplies far from home, motorists curtailed long-distance driving. This meant that gasoline became plentiful in most rural and resort areas, while shortages became increasingly concentrated in urban areas.

CONCLUSION

Efficient allocation involves marginal adjustments among competing users. An allocation system that establishes certain categories of priority users increases the cost of adjustment along certain margins (namely those between priority and nonpriority users), and thus is necessarily an inefficient allocation mechanism. Clear efficiency gains

42. Lane, *Mandatory Petroleum Regulations*, pp. 47-50.

could have been achieved had gasoline been reallocated from priority to nonpriority uses. Yet these adjustments could not be made because the allocation system itself made it too costly to do so. This inefficiency was at the heart of the disastrous experience with the mandatory allocation system in both 1973–74 and 1979. That proposals to reinstitute mandatory allocations precisely when they would do the most harm (i.e., in a period of supply disruption) should still be taken seriously—insofar as passage by both houses of Congress implies that a proposal has been taken seriously—as late as 1981 suggests either continuing confusion about the effects of the mandatory allocations or that priority users were able to use the political system to effect substantial wealth transfers in their own favor.

Despite the monumental failure of price controls and allocations and their direct responsibility for gasoline shortages, blame for the shortages was conveniently placed on an inconsequential oil embargo, on the greedy machinations of the oil industry, or on a nebulous energy crisis. An amusing and instructive example of such thinking is provided by the report of the House Interstate Commerce Subcommittee on EPAA, which Congress eventually enacted. The report stated in part:

> This country now has actual and imminent shortages in crude oil, residual fuel oil, and refined petroleum products. And whatever their origins, the Committee finds that these shortages are real, severe, and cannot be dealt with through reliance on the free market. . . .
>
> Some have suggested that the gasoline shortage has been contrived by the major oil companies to purge from the business their only significant competitors, the independent nonbranded dealers. The Committee does not have the means of assessing the truth of that allegation, but whether intended or not this clearly has been the result. The Committee firmly believes that we should not allow shortages of gasoline or, for that matter, shortages of crude oil, residual oil, or other refined petroleum products to cause shortages of competition. The self-regulating laws of supply and demand are not currently operating in the petroleum market. It is imperative that the federal government now accept its responsibility to intervene in these markets to preserve competition. Witnesses have testified that if we fail to do so, the cost to the consumers will ultimately be measured in many millions of dollars.[43]

43. House Report No. 93-531, 93rd Cong. 1st Sess., 1973, U.S. code Cong. and Admin. News, p. 2586.

While professing its concern over the existence of shortages, the Committee ignored any possible link between the controls that had already been enacted and the failure of "the self-regulating laws of supply and demand" to operate. Instead, the Committee found that the shortages could not be dealt with "through reliance on the free market," as if a free market had been allowed to function while the shortages were occurring. It went on to suggest further government intervention in petroleum markets (designed to favor various special interests) that only exacerbated the shortages and ensured their perpetuation. But our amusement at the irony of the Committee report will be somewhat tempered when we recall that the confusion it betrays and the special interests it serves still exert a potent influence on the shaping of public policy.

Chapter 6

PRICE CONTROLS ON GASOLINE AND OTHER REFINED PRODUCTS AFTER THE EMBARGO

"Do you feel the allocation formulas are hurting the Northeast?" Carter asked. "No sir," replied Schlesinger, "all the formulas did was help the West." He continued: "The effect is to . . ." Carter interrupted: "Put the gasoline where the cars are?" "No," replied Schlesinger. "What it does is to put the gasoline where the cars are not."*

MARKET-CLEARING PRICES UNDER CONTROLS

Understanding how controls created shortages is easy, though some seem to have acquired immunity to such understanding. What is not so easy to see is how controls could have been in effect for almost a decade without creating shortages for more than perhaps seven or eight months of that entire period.[1] We can formulate the problem as follows: What mechanism (or mechanisms) permitted the prices of gasoline and other refined products to attain market-clearing levels despite the controls to which they were subject? Another closely

*The epigraph to this chapter is an account of a conversation between President Carter and Secretary of Energy James Schlesinger conducted in the presence of reporters in July 1979 and reported in *Time*, 16 July 1979, p. 10.

1. The analysis of this section draws heavily on Harvey and Roush, *Petroleum Regulations*, which contains a more rigorous and complete analysis than I have undertaken here.

related problem that we must also look into is what other consequences, if any, were entailed by the operation of this, or these, mechanisms. Before we can tackle these problems, however, I must first describe in a simplified way the basic features of the controls on gasoline prices.

How Ceiling Prices for Gasoline Were Determined

Legal maximum prices that refiners could set for gasoline were determined by the following principle: A refiner could charge no more than the price charged on May 15, 1973, plus the refiner's subsequent increases in its average (per unit sales) costs.[2] Costs of labor and materials (like crude oil) could be passed through in the refiners' ceiling price. Some of the refiner's capital costs, such as depreciation and interest payments on debt, could also be passed through. But other costs, in particular the opportunity costs of equity capital, could not be passed through.

A brief note is in order concerning the difference between the capital costs that could be passed through and those that could not under the controls. Suppose a refiner's assets consist of $1,000 in plant and equipment and a $200 inventory of products and raw materials. The total value of assets is $1,200. The firm could have acquired these assets either with capital invested by its shareholders or with the funds obtained by borrowing. Suppose shareholders initially invested $600, and the remaining $600 was borrowed. Thus the firm has liabilities of $600 and a net worth of $600. Say the interest rate is 10 percent; then the cost to the firm of servicing its debt is $60 per year. Under the controls, such interest payments were considered allowable costs that could be counted in the calculation of the refiner's ceiling price. However, the owners of the firm would also bear an implicit cost of $60 per year in the interest foregone by keeping $600 invested in the firm rather than buying bonds yielding 10 percent interest. This implicit cost, which I have called the opportunity cost of equity capital, could not be counted in computing the refiner's ceiling price. As a consequence, controls provided an incentive to switch from equity to debt financing. However, since it is costly to increase debt-to-equity ratios (since borrowing costs rise as debt-equity ratios rise), the exclusion of equity capital costs dis-

2. Harvey and Roush, *Petroleum Regulations*, p. 1.

couraged the use of capital by firms subject to the regulations. Nor was this simply an oversight in the regulations that could have been corrected because in practice it is not possible to measure changes in implicit capital costs.

I must also mention several changes that occurred in the framework of the price control regulations for gasoline and refined products that were established under phase IV and codified in EPAA. In 1975 Congress passed the Emergency Petroleum Conservation Act (EPCA), which revised and continued the price controls and mandatory allocation provisions of EPAA. EPCA gave the President authority to terminate price controls and allocation for any refined product if he found that the product was not in short supply and if Congress did not veto his action. In 1976 President Ford decontrolled middle distillates, residual fuel oil, and a number of minor products. In 1979 President Carter decontrolled jet fuel and aviation gasoline, and in 1980 he decontrolled all other refined products except motor gasoline and propane, which were finally decontrolled by President Reagan on January 28, 1981.[3]

A fundamental problem in administering the controls on refined products was how to allocate the common costs of refining to any particular product. Refiners were allowed to allocate costs among refined products in any fashion they chose, subject to the condition that the sum of the shares assigned to the different products not exceed their volumetric share in total refinery output. As products other than gasoline were decontrolled, this posed a problem inasmuch as refiners had chosen to allocate a larger fraction of refining costs to gasoline than its volumetric share of output. This reflected the fact that various investments are required in refining processes in order to increase the relative share of gasoline in refinery output. Thus the decontrol of other products had a tendency to make the ceilings on gasoline prices more restrictive. To restore the incentive for gasoline production, the Department of Energy (DOE), in March 1979, allowed refiners to assign a percentage of the common refining costs greater than its pro rata volumetric share of refining output to gasoline.[4] This change in the regulations was called the "tilt" in favor of gasoline.

3. Lane, *Mandatory Petroleum Regulations*, pp. 126–27.

4. See Harvey and Roush, *Petroleum Regulations*, pp. 87–88. Refiners were allowed to take 110 percent of the operating costs attributable to gasoline on a pro rata volumetric

I now want to summarize the essential features of the price control regulations for gasoline, which will involve a bit of tedious notation and a rather complex formula for the calculation of a refiner's ceiling price for gasoline. I trust that the tedium will be balanced by the resulting increase in the power of our analytical apparatus, which will enable us to explain how controls were consistent with market clearing even when price ceilings were not above market-clearing levels. I shall use the following symbolic notation.

CP = a refiner's ceiling price for gasoline in the current period

BP = a refiner's base-period price for gasoline

QM = the amount of gasoline sold by a refiner in the current period

QR = the amount of products, measured volumetrically, produced by a refiner in the current period

QP = the amount of gasoline purchased for resale by a refiner in the current period

G = the volumetric share of gasoline in a refiner's total output of products

VRC = a refiner's variable refining costs in the current period

$ACRC$ = a refiner's allowable capital costs incurred in refining during the current period

VMC = a refiner's variable marketing costs in the current period

$ACMC$ = a refiner's allowable capital costs incurred in marketing during the current period

PP = a refiner's cost of purchasing gasoline for resale in the current period

When any of these variables are starred, they represent the values taken on by these variables in the base period. Note also that the total amount of gasoline produced by a refiner in the current period equals the volumetric fraction of gasoline in the refiner's total output (G) times the total output of the refiner in the current period

basis. The percentage of allowable nonproduct costs (e.g., interest payments and depreciation) they were allowed to take was given by the expression

$$2G - G^2$$

where G represents the volumetric share of gasoline in a refiner's output.

(QR) or in our notation, GQR. QM must therefore be equal to the sum of GQR and QP.

The ceiling price of gasoline to which a refiner would be subject under the controls would be calculated according to the following formula:

$$CP = BP + \{GQR\,[(VRC + ACRC)/QR - (VRC^* + ACRC^*)/QR\,] + QP\,(PP - PP^*)\}\,(1/QM) + (VMC + ACMC)\,(QM - VCMC^* + ACMC^*)\,QM^*.$$ [5]

The term $(VRC + ACRC)/QR$ represents average allowable refining costs in the current period, including both variable and allowable fixed capital costs. The same term with starred variables represents average allowable refining costs in the base period. Taking the difference between the two gives us the increase in the refiner's average allowable refining costs since the base period. If we multiply this increase by the amount of gasoline refined in the current period (GQR), we get the total increase in the refiner's allowable refining costs legally attributable to gasoline production. Dividing by the total amount of gasoline sold by the refiner (QM), we derive the increase in refining costs averaged over the amount of gasoline sold by the refiner in the current period.

Similarly $(PP - PP^*)$ represents the increase in the price the refiner pays for gasoline between the base period and the current period. Multiplying by the amount of gasoline purchased (QP) gives us the total increase in the cost of purchasing gasoline since the base period. Dividing by the total amount of gasoline sold (QM) gives us the increase in the cost of purchasing gasoline averaged over the total amount of gasoline currently sold. Thus the expression multiplied by $(1/QM)$ represents the allowable increase in the refiner's average cost of refining and purchasing gasoline.

The difference between the last two terms of the equation represents the increase in the refiner's average allowable cost of marketing gasoline since the base period. Adding the allowable increases in the refiner's average costs of refining, purchasing, and marketing gasoline to the refiner's base period price gives us the refiner's current ceiling price.

5. This equation is, except for changes in notation, the same as equation (3-4) in Harvey and Roush, *Petroleum Regulations*, p. 60. My equation includes no tilt. For the equation including the tilt see Harvey and Roush, p. 88, equation (4-2).

A numerical example may help to show how ceiling prices were calculated and how they could be altered by the behavior of individual refiners. Let us suppose that in the base period the price of gasoline sold by a refiner was 50¢ per gallon. Suppose that this refiner sold 100 gallons of gasoline, of which the refiner produced 50 in his own refinery and purchased 50 in the spot market. Assume that the refiner paid 30¢ per gallon for the gasoline it purchased in the base period and now pays 40¢ per gallon. Let us also assume that half of the refiner's total output was gasoline, so that its total output of refined products was 100 gallons. Further, say that the refiner's base-period variable refining costs plus the allowable capital costs incurred in refining equaled \$30 and that its variable marketing costs plus its allowable capital costs incurred in marketing equaled \$8. In the current period let the former cost be \$40 and the latter \$10, and assume that there has been no change in the refiner's total output of gasoline and other refined products or in its purchase of gasoline since the base period. Substituting these values into the above equation, we find the refiner's ceiling price:

$$\$.50/\text{gal} + \Big\{ 50/\text{gal}\,[\$40/100\text{gal} - \$30/100\text{gal}] + 50\text{gal}\,(\$.40/\text{gal} - \$.30/\text{gal}) \Big\}\,(1/100\text{gal}) + \$10/100\text{gal} - \$8/100\text{gal} = \$.62/\text{gal}$$

The refiner's average variable refining costs plus its average allowable capital costs in refining have increased by 5¢ per gallon sold; its average cost of purchasing gasoline has increased by another 5¢ per gallon sold; its average variable marketing costs plus its allowable capital costs in marketing have risen by another 2¢ per gallon. Together these increases account for the 12¢-per-gallon increase in the current ceiling price over the base-period price.

Complicated as it may seem, the above description is actually a simplification of the way that ceiling prices were calculated. Refiners did not face just one ceiling price for gasoline, either, as each grade of gasoline sold had its own ceiling price. A source of much greater complexity was the fact that refiners normally sell gasoline to different customers at different prices depending on the services provided and the transportation costs incurred. Refiners were, therefore, allowed to group customers charged comparable prices into separate "classes of purchasers."[6] The refiner would then calculate a separate

6. The term "class of purchaser" was defined as "purchasers to whom a person has charged a comparable price for a comparable property and service pursuant to customary

ceiling price for each class of purchaser. Sales within each class were subject to the corresponding ceiling, and the price differentials between classes could not be changed from those prevailing in the base period.[7]

Whenever a refiner was selling below its ceiling prices within each of its classes of purchasers, it was allowed to "bank" its unrecovered costs. The "equal application" rule provided that a refiner might only bank costs to the extent of the minimum difference between selling price and ceiling price among all its classes of purchasers. The question that immediately presents itself is why a refiner would choose to charge less than its ceiling price in any, let alone all, of its markets. At least one of the answers is that the market-clearing price enforced by competition was often a more severe constraint than the ceiling prices established by the controls.

Banked costs could subsequently be drawn upon when ceiling prices did become binding and were below market-clearing levels. Thus it might appear that if a refiner had positive banked costs, ceiling prices could not be binding. This would probably be an unwarranted inference for several reasons. First, refiners could not determine precisely what their ceiling prices were at any time since the ceiling prices would depend on costs the refiner was incurring whose exact magnitude could only be ascertained later. Thus unless a refiner provided some margin for error, it could easily exceed its ceiling price inadvertently. As long as the refiner had positive banked costs, no great harm would result from a temporary breach of its price ceiling. But if a refiner had negative banked costs for two consecutive months, it was subject to penalties for violating the price regulations. There was, therefore, an incentive for refiners to build up and maintain banked costs even when ceilings were binding.

Second, some refiners may have been constrained while others were not. That the industry as a whole had accumulated banked costs does not demonstrate that no refiners were being constrained by the price ceilings in any of their transactions.[8]

---

price differentials between those purchasers and other purchasers." 10 C.F.R. Section 212,31 (1980), quoted in Lane, *Mandatory Petroleum Regulations*, p. 149, note 241.

7. "The intent of this regulatory framework was to preserve the 'customary price differentials' between purchasers, which, in fact, serve to define 'class of purchaser' in Section 212,31, by requiring the application of equal amounts of increased product costs to the May 15, 1973, selling price of each class of purchaser." 39 Fed. Reg. 32,307 (1974), quoted in Lane, *Mandatory Petroleum Regulations*, p. 150. Also see Harvey and Roush, *Petroleum Regulations*, pp. 112–15, and U.S. Department of Energy, *State of Competition*, p. 223.

8. See Harvey and Roush, *Petroleum Regulations*, pp. 185–89.

How Ceiling Prices Could Be Above Market-Clearing Prices

The presumption that shortages are impossible without price controls is so overwhelming that when an economist observes price controls on any commodity for very long without also observing shortages, he or she will generally conclude that the price controls were above market-clearing levels. Many economists have made this claim about the controls on gasoline and other refined products between 1974 and 1979.[9] Presently I will discuss reasons for supposing that controls were not binding in many cases, but a mechanism by which refiners could raise binding ceiling prices to market-clearing levels also existed. The two possibilities are not identical. If market clearing was achieved only because price ceilings were not binding, then the ceilings were redundant and had no impact on anyone's behavior. But if price ceilings were binding, and refiners could still raise their ceiling prices to market-clearing levels, then the process by which prices were raised presumably had further repercussions, which must be explored. But first I should like to discuss the reasons that ceiling prices might not have been binding.[10]

Recall that price ceilings were determined by allowing increases in out-of-pocket costs per unit sales to be added to the base-period price. However, the market-clearing price would not necessarily rise as much as the ceiling price in the event of an increase in certain input prices. The reason is that increased costs arising from increased crude oil prices may tend to depress the prices of cooperating inputs, thereby limiting the increase in product prices. If the costs of these cooperating inputs were not counted in the calculation of ceiling prices, the ceiling prices would only reflect the higher cost of using crude oil but not the lower cost of using other inputs. If such were the case, the ceiling price for gasoline could surpass the market price.

Why would the costs of cooperating inputs not counted in calculating ceiling prices for gasoline have fallen as the price of crude oil

9. For example, Phelps and Smith, *Dilemma of Decontrol*, pp. 24–29, and Kalt, *Oil Price Regulations*, pp. 37–44. Kalt and Phelps and Smith infer that controls were not binding during this period from the fact that banked costs were positive. The above discussion indicates that such an inference is not necessarily warranted.

10. The following discussion is based on the analysis in Phelps and Smith, *Dilemma of Decontrol*, pp. 11–23.

rose? Let us suppose that crude oil and refining inputs are combined to produce gasoline in fixed proportions so that a refiner could not substitute additional refining inputs for crude oil without suffering a loss of gasoline output. What if the price of crude oil were to rise? Insofar as the increased price of crude were passed through to the price of gasoline, the sales of gasoline would fall and so necessarily would the amount of crude oil refined. Since crude oil and refining inputs are combined in fixed proportions, the demand for refining inputs would fall correspondingly.

Whether the price of refining inputs would also fall depends on the opportunities available to refining inputs in other activities. If opportunities elsewhere offered comparable remuneration, then their price in refining could not be reduced significantly. While one can well imagine refinery workers taking their services elsewhere rather than accepting lower wages in the refining industry, it is a bit harder to imagine the owner of a refinery trying to use it as anything other than a refinery. The only alternative to using a refinery for refining is not to use it at all. Thus the owner of a refinery must be prepared to allow the remuneration received for its services to fall at least to zero in the event of a sharp enough reduction in demand for refining services.[11]

A reduction in the cost of capital specialized to refining would partially offset the increase in the cost of crude oil, but the reduction in the cost of refining capital was not reflected in the ceiling price of gasoline because refining equipment is owned by the refiners. The cost of refining equipment is the opportunity cost of its continued ownership by the refiners, which cost obviously falls as the value of the equipment falls. Such implicit costs were excluded from the calculation of the ceiling price, so the ceiling price reflected only the increase in crude oil prices, not the reduction in the cost of the refinery equipment.

There seems to be good reason for supposing that ceiling prices for gasoline were often not binding. Yet the tremendous complexity of the market, the heterogeneity of refining equipment, and the widely differing ceiling prices to which individual refiners were subject should caution us against assuming that ceiling prices were never

11. In the short run at least, the refiner might allow the remuneration to become negative (i.e., to operate at a loss) if neither shutting down the refinery temporarily nor abandoning it permanently is costless.

binding. It is unlikely that all firms were always and everywhere selling gasoline for less than their ceiling prices; one might then conclude that refiners were charging less than market-clearing prices. But as I suggested before there were mechanisms by which refiners could raise their price ceilings to market-clearing levels (which is different from saying that price ceilings were not binding).

How Ceiling Prices Could Be Raised to Market-Clearing Levels

Let us return to the numerical example above.[12] The ceiling price we calculated was 62¢ per gallon. Suppose the market-clearing price is above that, say 65¢ per gallon. Note first of all that if—in the absence of controls—the refiner had been selling gasoline for 62¢ per gallon and could purchase gasoline for 40¢ per gallon, then (if we ignore for a moment some complicating factors) the marginal cost of marketing the gasoline must have been 22¢ per gallon. If it were only, say, 20¢ per gallon, the refiner could have bought the gasoline for 40¢ per gallon, incurred the 20¢ per gallon cost of marketing the additional gasoline, resold it at 62¢ per gallon, and made 2¢ clear profit on each gallon purchased and resold. If the marginal marketing cost were greater than 22¢ per gallon, say 40¢ per gallon, then it clearly would have been worthwhile for the refiner to reduce its purchases of gasoline for resale since, at the margin, it would have been losing 18¢ on a gallon bought and resold. Presumably, the marginal cost of marketing increases with the amount marketed, so that if marginal cost is less than the difference in the buying and selling prices, increasing purchases eventually brings the marginal cost into equality with the spread. The relationship between marginal cost, sales, and the spread between the buying and selling prices is shown in Figure 6-1. The intersection of the marginal cost curve with the horizontal line representing the price spread indicates the optimal level of sales for the refiner at retail.

Under controls, however, even if the refiner's marginal cost of marketing gasoline were somewhat more than 22¢ per gallon, it

12. The argument of this section is essentially a reformulation of the analysis of ceilings on gasoline prices presented by Harvey and Roush, *Petroleum Regulations*, especially Chapter 3.

Figure 6-1. Optimal Level of Sales by a Gasoline Marketer.

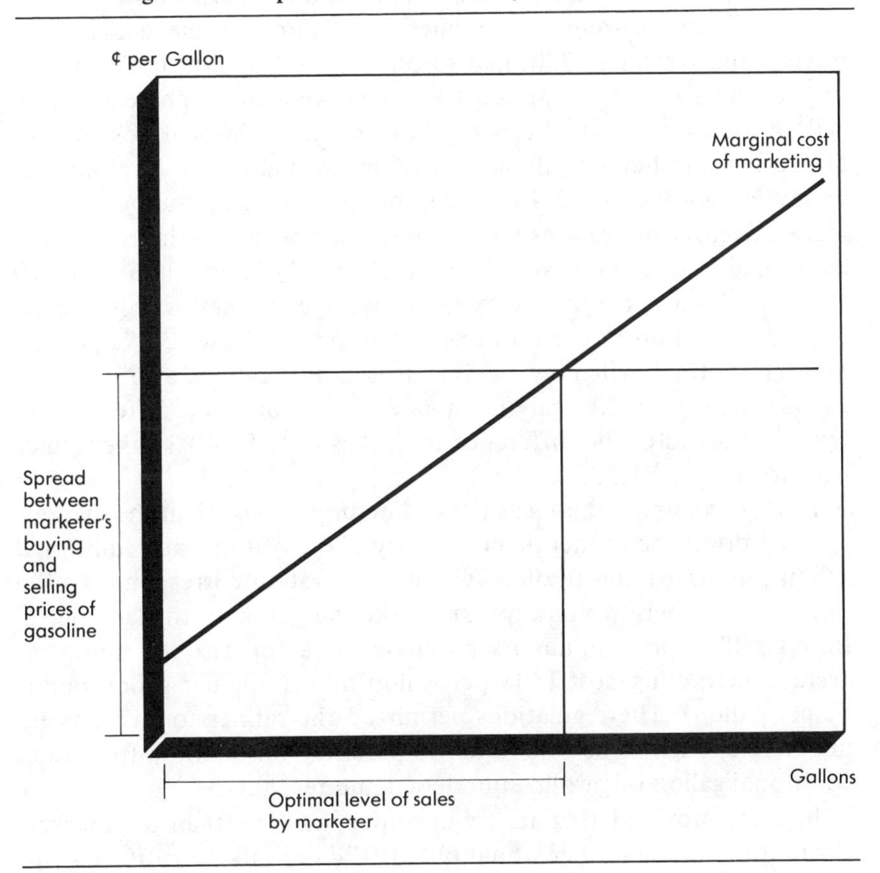

would pay the refiner to purchase and resell additional gasoline because the resulting increase in its average costs would lift its ceiling price. In our example, since the *average* marketing cost is only 10¢ per gallon, while the *marginal* marketing cost is 22¢ per gallon, any increase in the volume of gasoline marketed must raise the *average* marketing cost.[13] Hence, the refiner would receive more than 22¢ additional revenue from each gallon of gasoline bought and resold, so

13. In our case we have a total cost as a function of quantity. From this we derive marginal cost and average cost as functions of quantity. The marginal cost represents the rate of change of total cost at any quantity, and average cost represents total cost divided by quantity. If marginal cost exceeds average, it must raise the average; and if marginal is less than average, it must lower the average.

the refiner would be willing to incur more than 22¢ per gallon in additional marketing costs to market the additional gasoline.

Let us work through our numerical example to make clear how buying and reselling additional gasoline would be profitable for our hypothetical refiner. Suppose the refiner were to purchase 10 more gallons of gasoline at 40¢ per gallon, or an additional cost of $4. Marketing another 10 gallons of gasoline would cost the refiner, let us say, an additional $2.40, or 24¢ per gallon. (I am supposing that marginal costs increase as volume increases, which is the reason the additional marketing cost is not $2.20, or 22¢ per gallon times 10 gallons.) Owing to the increase in average marketing costs, now 11.27¢ per gallon,[14] the refiner's ceiling price is now 63.27¢ per gallon. The refiner will now be able to sell 110 gallons at 63.27¢ per gallon instead of 100 gallons at 62¢ per gallon. The difference in revenue is $7.60. The difference in cost is only $6.40, so the refiner has profited by $1.20.

In this example, so long as the ceiling price is less than the market-clearing price, the refiner makes exactly 12¢ profit on each additional gallon purchased and resold. Twelve ¢ per gallon corresponds to the refiner's base-period markup for marketing gasoline, that is, the refiner's selling price minus its purchase price for gasoline minus its *average* marketing cost (50¢ per gallon minus 30¢ per gallon minus 8¢ per gallon). The regulations permitted the refiner to raise its ceiling price enough to be able to derive precisely this markup from each additional gallon of gasoline purchased and resold.

In fact, provided that its ceiling price was less than the market-clearing price, the refiner would obtain 12¢ per gallon profit–regardless of the price it paid–from any additional gasoline it bought and resold. A lower price was preferable to a higher one because the lower the price, the more gasoline the refiner could purchase and resell before raising its ceiling price to the market-clearing level. Nevertheless, if supplies were tight and spot prices rising, refiners with ceiling prices below market-clearing levels (they were most likely to be below market-clearing levels in just such periods) would not be deterred from paying prices much higher than they could have profitably paid in the absence of controls.

Another numerical example will clarify this point. Suppose that at 40¢ per gallon the refiner can obtain only 50 gallons, but that additional supplies of gasoline are available at 50¢ per gallon. Given that

14. The figure 11.27¢ per gallon equals $12.40 divided by 110 gallons.

the refiner must incur a marginal cost of at least 22¢ per gallon to market gasoline, it might appear unprofitable for the refiner to purchase gasoline for 50¢ for resale when its ceiling price is only 62¢ per gallon, but this is actually not the case. Say the refiner goes ahead and buys 10 gallons at 50¢ per gallon in addition to the 50 gallons it was already buying at 40¢ per gallon. The refiner's ceiling price would then work out to 64.18¢ per gallon.[15] Its revenue from selling 100 gallons at this price would be $70.60. This is $8.60 more than it would collect selling 100 gallons at 62¢ per gallon. The refiner's costs would rise by $7.40–$5 to buy 10 gallons at 50¢ per gallon and $2.40 in additional marketing costs for the extra 10 gallons. The refiner is left with an increase in profit of $1.20 from buying and selling the additional 10 gallons–the same 12¢ per gallon that the refiner made in the previous example. The only difference here is that refiner's ceiling price rises more quickly when buying additional gasoline for 50¢ per gallon than when buying it for 40¢ per gallon. Since the ceiling price rises more quickly the higher the price the refiner pays for additional supplies, the amount of gasoline the refiner purchases before raising its ceiling price to market-clearing levels (and hence the amount of increased profits accumulated) is less the higher the price the refiner must pay for additional supplies.

Although it was more profitable for a refiner constrained by price ceilings to pay a low price than a high price for more gasoline, buying some gasoline, *whatever the price*, was clearly more profitable than buying none. In effect, refiners received a subsidy (in the form of an increased ceiling price) for any gasoline they could purchase and resell so long as their ceiling prices remained below market-clearing levels, no matter how much they paid for the gasoline. They were encouraged to buy products for resale at prices far higher than they would have been willing to pay without the subsidy (i.e. without the price ceilings).[16]

These incentives became most perverse precisely in times of shortages when prices in spot markets were rising rapidly and ceiling prices were lagging behind. Refiners found themselves constrained by price

15. The figure 64.18¢ per gallon is derived as follows: The base price, 50¢ per gallon, plus the increase in average variable plus allowable capital costs ($5 per 110 gallons), plus the increase in the average cost of purchasing gasoline (50 gallons at 40¢ per gallon plus 10 gallons at 50¢ per gallon, divided by 100 gallons minus 30¢ per gallon), plus the increase in average marketing costs (11.27¢ per gallon minus 8¢ per gallon). After performing all these calculations, we arrive at a ceiling price of 64.18¢ per gallon.

16. Lane, *Mandatory Petroleum Regulations*, pp. 140-42.

ceilings that were below market-clearing levels. Seeking to raise their ceiling prices to market-clearing levels, refiners had the incentive to enter the spot market and obtain marginal supplies at prices above market-clearing levels because the added costs would be recovered as their ceiling prices rose accordingly. Competition for additional supplies that could be used to raise ceiling prices helped to generate the rapid increases in prices that characterized the periods of severe shortages in 1973–74 and 1979.[17]

This process may well have contributed to expectations of further rapid increases in petroleum prices that in turn encouraged sellers to withhold supplies in anticipation of these increases. Without such expectations it is possible that the general level of petroleum prices would not have risen so high either in 1973–74 or 1979. Also since increases in ceiling prices lagged behind increases in the general level of petroleum prices, rapid increases in crude oil prices could, paradoxically, trigger short-term shortages. Thus, under price controls, the process of rising prices and shortages had a certain self-reinforcing tendency.[18]

I want to mention one other mechanism by which ceiling prices could be raised to market-clearing levels. Even though refiner-marketers could raise their ceiling prices through the mechanisms I have described, the number of different ceiling prices to which their various transactions were subject and the fact that differentials between these prices were frozen by the regulations made it impossible for any refiner to set its ceilings equal to market-clearing prices in all markets. In some markets a refiner's price ceilings might remain below market-clearing levels, while in others the ceilings could be above market-clearing levels.[19] This accounts for the extraordinary price differentials observed in many product markets while the con-

17. See my discussion of the retail pricing of natural gas in Chapter 10 and Appendix B.

18. In explaining the mechanism by which refiners could raise their ceiling prices to market-clearing levels, I have concentrated on the purchase of gasoline for resale. But a refiner could also raise its ceiling prices by refining additional crude oil. Since doing so increases the refiner's average refining costs, the refiner would increase its ceiling price and could sell all of its output at a higher price. The difficulty with increasing refinery output is that costs would necessarily be incurred for the production of decontrolled products from which there would be no extra return in the form of a higher ceiling price. Although it is possible that some refiners may have increased their ceiling prices for gasoline by purchasing and refining additional quantities of crude oil, the purchase and resale of gasoline would appear to have been the more reliable method. See Harvey and Roush, *Petroleum Regulations*, pp. 73–79.

19. See Harvey and Roush, *Petroleum Regulations*, pp. 172–76.

trols were in effect. Such differentials could not have persisted in the absence of controls.

Where refiners were selling to customers at less than market-clearing prices, there was an incentive for the customers to resell the products they bought at the market-clearing price and pocket the difference. Suppose a refiner was selling to some of its jobbers or wholesale customers below market-clearing prices. Since these purchasers were themselves subject to price ceilings that limited the markups they could take, they could not necessarily capture the windfalls associated with the low cost of obtaining supplies. To believe, however, that no one would capture these windfalls before the gasoline reached consumers is to betray a quaint naivete concerning the motives of most businesspeople or else a fatuous underestimation of their ingenuity.

To capitalize on these profit opportunities, intermediaries would purchase and resell products among themselves until ceiling prices were raised to market-clearing levels. These sequences of exchanges among intermediaries came to be known as "daisy chains."[20] Since products would often change hands several times between the refiner and the final consumer, it was not easy for outsiders to distinguish between legitimate transactions and mere links in the daisy chain. Furthermore, because the sums of money involved in the gasoline daisy chain were very much smaller than those in the crude oil market, they attracted relatively little attention from the press or government officials. As one of the drafters of the price control regulations over the oil industry put it:

> The reseller who was vertically integrated and performed all of the marketing and distribution activities from the point the product left the refinery to the time it reached the service station had a radically different financial structure than the reseller who was an individual broker operating over his telephone. The respective margins of these two different kinds of resellers were simply not comparable.[21]

The wide variance in resellers' margins offered them ample opportunity to engage in daisy-chain transactions with little or no risk of detection. These transactions provided no real service to consumers

20. See Lane, *Mandatory Petroleum Regulations*, pp. 206-07. Also see Cook, "Daisy Chain Scandal," for a rather sensationalized account of the daisy chains operating in the crude oil market. The daisy chains and the role of regulations in promoting them are described at 45 Fed. Reg. 8,095 (1980) and 45 Fed. Reg. 74,432 (1980).

21. Owens, "Petroleum Price Controls," pp. 1268-69.

or anyone else. Whatever costs were incurred in order to execute them were, in and of themselves, a pure waste. However, by enabling prices to rise to market-clearing levels, the daisy chains did help to avoid even greater wastes resulting from actual shortages.

The Inefficiencies of Price Ceilings Even When Raised to Market-Clearing Levels

Having shown how refiners and intermediaries were able to raise binding price ceilings to market-clearing levels, I want to round out the discussion by enumerating some of the inefficiencies that arose in the process of raising ceiling prices to market-clearing levels.

1. Under controls, certain costs (labor, materials, and interest payments, for example) could be recovered through increased ceiling prices, but other costs (primarily implicit capital costs) could not. The unequal treatment of different kinds of costs created a bias against using inputs of the latter type. Since the implicit costs of investment in the expansion or modernization of refinery capacity were not reflected in ceiling prices, refiners avoided making such investments under controls. Modernization of refinery capacity would have increased the yields of the more valuable, lighter refined products such as gasoline and middle distillates (heating oil and diesel fuel). It would also have allowed refiners to reduce their use of the lighter, lower sulfur content grades of crude oil, which have been making up a declining portion of the world's crude output and commanding rising premiums over the heavier grades. Thus excluding implicit capital costs from the calculation of ceiling prices tended to increase gasoline prices in the long run by reducing the output of gasoline relative to what it would have been without the exclusion. In addition, the small refiner bias of the entitlements program encouraged investment in small, inefficient, "tea-kettle" refineries that produce little or no light products.
2. The short-run effect of controls on gasoline prices at the refinery gate is ambiguous. The direction of the effect depends on whether constrained refiners increased the amount of crude oil they were refining in order to raise their ceilings. If gasoline were

the only refined product, or if all products were subject to price ceilings and common costs could be allocated among products at the discretion of refiners, then constrained refiners certainly would have done so. But under partial decontrol, the return from refining gasoline was limited by the volumetric restriction on the assignment of common refining costs. As a result it is not obvious whether it was profitable for constrained refiners to increase their refinery output of gasoline.[22] The other method of raising the ceiling price of gasoline—purchasing gasoline upstream and reselling it downstream—would seem to have been the more profitable strategy under partial decontrol. If so, controls on gasoline prices (when other products were decontrolled) may have reduced the production of gasoline by constrained refiners who would have purchased gasoline for resale instead of producing it. If that indeed is how constrained refiners responded to controls, then total refinery output of gasoline would have fallen. Unless the consequent loss of domestic refinery output of gasoline was made up for by increased imports, it would appear that controls on gasoline prices along with partial decontrol of other product prices may have increased the price of gasoline. This must remain a rather tentative conclusion. But certainly the long-run effect mentioned above makes it probable that, over time if not immediately, the controls did raise prices for gasoline consumers.

3. One cannot, however, conclusively rule out the possibility that at least some constrained refiners were—even under partial decontrol—induced to refine additional gasoline as a way of increasing their ceiling prices. In the absence of controls, a refiner would produce an amount of gasoline such that its marginal refining cost of gasoline was just equal to the price of gasoline at the refinery gate. However, if a constrained refiner found it profitable to raise its ceiling price by increasing its output of gasoline, this refiner would produce an amount of gasoline such that its marginal cost of refining gasoline exceeded the price of gasoline at the refinery gate. The difference would correspond to the additional revenue obtained from the increased ceiling price. Now if some refiners were induced to raise their ceiling prices by increasing their refinery output of gasoline and others were not,

22. See footnote 18 above.

then some refiners must have been operating with higher marginal costs than others. But a necessary condition for the efficient provision of a product by several suppliers is that the marginal costs of all suppliers be equal. If they aren't, savings could be achieved by reducing the output of suppliers with high marginal costs and increasing the output of suppliers with low marginal costs. So if the marginal refining costs of refiners were unequal, the production of gasoline was not efficiently apportioned among them. Those with high marginal costs were producing too much gasoline, and those with low marginal costs too little.

4. Since refiners constrained by price ceilings could use purchases of gasoline to raise their ceilings, the net value to them of the gasoline purchased was greater than the price for which it could be resold. Constrained integrated refiners thus expanded their marketing operations so that marginal marketing costs exceeded the spread between the buying and selling prices. This implied (unintentional) predatory competition against unconstrained nonintegrated and integrated marketers by constrained refiner-marketers,[23] a practice that was inefficient as well as inequitable. Since constrained refiner-marketers were receiving an implicit subsidy on their marketing operations that by its nature was unavailable to any unconstrained marketer, constrained refiner-marketers must have had higher marginal costs than unconstrained marketers. Thus, constrained refiner-marketers marketed too much gasoline, and unconstrained marketers, too little. This inefficiency was at least partially borne by consumers who paid increased prices for gasoline because of an inefficient distribution network.

5. When refiners were unable to raise the ceiling prices constraining their sales in particular markets to market-clearing levels, an incentive arose to enter into uneconomic transactions merely for the purpose of accumulating markups on the successive transactions that would raise the applicable ceiling price to a market-clearing level. Insofar as these daisy-chain exchanges used up any real resources, they constituted a waste attributable to the existence of the price ceilings.

23. Nonintegrated marketers, in principle, had a similar incentive to purchase and resell gasoline in order to raise their ceiling prices. But regulations on nonintegrated marketers were such that their price ceilings were much less likely to be binding than were the price ceilings of refiner-marketers.

6. During shortages, the price ceilings encouraged refiners and marketers to bid up the prices of additional supplies in order to raise their ceiling prices. Besides encouraging the importation of crude oil and products at a cost exceeding their value to consumers, the controls may have induced foreign producers to raise prices during shortages even more drastically than they otherwise could have. As will be shown in Chapter 10 and Appendix B, enforcing price ceilings based on average costs causes the marginal supplier to set a price higher than would have been possible if prices to consumers had corresponded to marginal cost.
7. The regulations governing price differentials reduced the competitiveness of retail and wholesale markets by making it more costly for refiners and intermediaries to adjust relative prices among different customers in response to changing market conditions. Suppose a refiner was selling at the same market-clearing price in two markets, A and B, and that the binding ceiling prices in both markets were just equal to the market-clearing prices. What if the market-clearing price in market A fell? Since a refiner would normally have to incur costs to raise the ceiling price to market-clearing levels, the refiner might well want to allow its ceiling price in market A to fall, say by reducing its market purchases of gasoline for resale in market A. But the equal application rule required that differentials between ceiling prices of different markets (classes of customers) remain the same as they were in the base period. So if the refiner were to reduce its ceiling price in market A to the new market-clearing level, it would have to reduce the ceiling price in market B below the market-clearing level. Just how a refiner would respond to this situation cannot be determined in general, but in many instances there would be an implied reluctance to initiate or follow price discounting in markets where to do so might, in the absence of controls, have been a profitable strategy.[24] Thus the rigidities inherent in controls were at least partially responsible during most of the seventies for the absence of the gasoline price wars that so characterized gasoline markets in the fifties and sixties. Their reemergence shortly after controls on gasoline prices were lifted is testimony to the anticompetitive effects of the controls.[25]

24. See Department of Energy, *State of Competition*, p. 223, footnote, Harvey and Roush, *Petroleum Regulations*, pp. 172–76, and Lane, *Mandatory Petroleum Regulations*, pp. 149–52.

25. *The Wall Street Journal*, 9 March 1982, p. 14.

It would, moreover, be a fundamental error to attribute the inefficiencies caused by the controls on the prices of gasoline and other products to inadequate design of the regulations or to think that distortions could have been avoided by better drafting of rules or more intelligent administration of the program. As Harvey and Roush put it:

> The cause of most of the inefficiencies . . . is the failure of the regulations to provide for marginal cost pricing. In most public utility situations, this failure is due to an inability to determine marginal cost or to implement the correct pricing scheme. In the case of the petroleum regulations, the problem is more fundamental. The imposition of what is essentially average-cost pricing is done intentionally to extract rents from the regulated firms.
>
> The adverse effects of DOE regulations cannot be dismissed, therefore, as being simply the result of poorly designed rules. They reflect general problems encountered when regulations are used to supplant market forces.[26]

THE RETURN OF GASOLINE SHORTAGES

By April 1974 ceilings on gasoline prices had risen enough to eliminate gasoline shortages. The conjunction of higher prices with the disappearance of shortages did not convince many people that the price controls had been responsible for the shortages because, although prices had risen, controls remained in effect. That price controls remained in force suggested to most people that gasoline prices would have been even higher in their absence. It was therefore the lifting of the Arab oil embargo in April that was credited with having eliminated the shortages.

For the next four years consumers enjoyed a respite from rapid price increases and supply disruptions. From the spring of 1974 to the fall of 1978, petroleum prices did not keep up with the rate of inflation. But the Islamic revolution in Iran that toppled the Shah in January 1979 and its aftermath of instability and violence so unsettled world markets that the price of crude oil rose almost continuously from January 1979 to March 1980. The doubling of oil prices during that period was occasioned not merely by the temporary suspension and permanent reduction of Iranian oil production, but also

26. Harvey and Roush, "Petroleum Price Regulations," p. 140, quoted in Lane, *Mandatory Petroleum Regulations*, note 2, p. 145.

by fears—exacerbated by the American hostages in Iran and the Soviet invasion of Afghanistan—that oil supplies from the Persian Gulf were becoming increasingly susceptible to interruption, owing to the instability of the oil-producing states or external aggression.

At first, domestic gasoline markets were not too seriously affected by the turmoil in world markets. Gasoline prices were able to rise substantially because the banked costs refiners had previously accumulated allowed them to raise their prices fairly quickly in response to tightening market conditions. But even though banked costs helped keep prices generally near market-clearing levels, rigidities in the price structure became more serious in tight-market conditions. Banked costs had to be applied equally to all classes of purchasers and, thus, could not be concentrated in markets with the most severe excess demands. As a result, scattered shortages surfaced late in 1978.[27]

Not until May, when serious shortages occurred in California, did the supply problems become unmanageable. When the shortages spread to many other parts of the country in June, it seemed that the repeated warnings of an eventual recurrence of the energy crisis had finally been confirmed. The havoc caused by the shortages from May through July 1979 incited public indignation and outrage against both the oil industry and the Carter administration. Since there was no oil embargo on which the shortages could be blamed, the public was more than ever inclined to blame the oil companies instead. Try as he would to take an antioil-industry stance, President Carter could not avoid being regarded as either too weak to resist the oil industry's greed or too ineffectual to thwart its conspiracies.

Facing an imminent collapse of such public confidence as he had maintained until then, President Carter retreated to Camp David in search of some way to dispel the public anger, distrust, and contempt that were undermining his administration fully a year and a half before the expiration of his term. After two weeks of almost mysterious seclusion, Mr. Carter emerged on July 15, 1979, to inform the American people that the crisis they faced was not—as they had mistakenly thought—one of trying to find gasoline with which to run their cars but one of the soul and the spirit. The crisis—characterized by the pursuit of narrow, selfish goals, a loss of national self-confidence, and a paralysis of public policy—had allowed the United

27. Lane, *Mandatory Petroleum Regulation*, p. 60.

States to become dangerously dependent on imported oil in general and OPEC oil in particular. That dependence, Mr. Carter observed in all seriousness, was "the direct cause of the long lines which have made millions of you spend aggravating hours waiting for gasoline." For good measure, Mr. Carter also ascribed rising inflation and growing unemployment to the dependence on imported oil. The energy crisis was real and worldwide, and it posed a clear and present danger to national security. But at bottom the President seemed to be saying that it was the American people who, because of their own greed and waywardness, were responsible for the crisis.

Mr. Carter's attempt to link gasoline shortages with defects in the national character did little to restore public confidence in his presidency. The implausibility of President Carter's interpretation coupled with his refusal to explain or his unwillingness to comprehend that the very price controls and allocations his administration was enforcing had precipitated the crisis only reinforced the inclination of the public to seek their own scapegoats for the shortages. Responding to widespread charges of a conspiracy by the oil industry to withhold supplies and create shortages, Mr. Carter ordered both the Department of Justice and the Department of Energy to investigate the role of the oil companies in the shortages. The reports of both departments refuted the charges of conspiracy and instead placed the primary responsibility for the shortages on controls over prices and the mandatory allocation of supplies.[28]

Indeed the Department of Justice report found that Mr. Carter's own voluntary wage-and-price guidelines had played an important part in causing the shortages. Mr. Carter had announced these guidelines on November 1, 1978, in an attempt to check a rapid and accelerating inflation. Under the guidelines, firms were admonished not to increase the weighted average of their product prices by more than (1) the rate of price increase during the applicable base period less a "deceleration percentage" or (2) 9.5 percent, whichever was less. Certain industries, including petroleum refining, were subject instead to a gross-margin test that limited increases in gross margins during the first year of the program to 6.5 percent. Since price ceilings under EPCA had already been lifted from many products, the voluntary guidelines, if observed, were the only constraint on price increases for these products. And although the guidelines were termed

28. U.S. Department of Justice, *Gasoline Shortage*, and U.S. Department of Energy, *Oil Supply Shortages.*

"voluntary," firms found by the Council of Wage and Price Stability to have violated the guidelines were to be subjected to official and public criticism for their misdeeds, and members of the public would be asked not to patronize them. Even more significant for the major oil companies, perhaps, was that violators were to be denied the right to bid on new leases to explore for oil and gas on federal lands.[29]

There is persuasive indirect evidence that the voluntary guidelines were often the binding constraint on refiners' prices in 1979. First of all, banked costs increased dramatically between December 1978 and July 1979, rising from $532 million in December 1978 to $1.11 billion in February 1979 and to $2.989 billion in July 1980.[30] Since one would expect refiners to draw down their banked costs in order to raise their prices during shortages, it seems anomalous that banked costs would have begun to increase rapidly while shortages were becoming increasingly serious. But if refiners were constrained by the voluntary guidelines, one can well understand why banked costs should have increased even during a period of shortage.

Second, shortages of diesel fuel, which had been decontrolled in 1976, began to occur in the spring of 1979. These shortages led to outbreaks of violence by truckers unable to obtain supplies of diesel fuel, just as similar ones had in 1973-74. Since it was manifestly not DOE price ceilings that were preventing diesel fuel prices from reaching market-clearing levels, the voluntary guidelines were the only possible obstacle to their having done so.[31]

Whether it was the guidelines or the price ceilings that were responsible, prices were held below market-clearing levels, and spot shortages began to appear by the end of 1978. In December 1978 Shell and Amoco had to begin allocating unleaded gasoline to their customers. In early 1979 other refiners began allocating products as well.[32]

Once a refiner found that the demands of all its base-period customers (to whom it was legally required to sell products) exceeded its available supply, it was required to place those customers on allo-

29. Lane, *Mandatory Petroleum Regulations*, pp. 57-58 and Department of Justice, *Gasoline Shortage*, pp. 12-13.

30. U.S. Dept. of Justice, *Gasoline Shortage*, p. 16.

31. Lane, *Mandatory Petroleum Regulations*, pp. 73-75. A third bit of evidence that it was the voluntary wage-price guidelines that constrained prices from reaching market-clearing levels is that price spreads for distillate in March and April 1979 were 3 to 7¢ per gallon, the same as for gasoline. Only if guidelines were binding on some refiners could such large spreads have persisted. See Lane, p. 58.

32. Lane, *Mandatory Petroleum Regulations*, p. 60.

cation. As they did in 1973-74, the allocation rules assigned gasoline to certain users according to politically determined criteria. Competition for supplies was thus transferred from the economic to the political marketplace. Instead of competing by offers of exchange, potential users had to compete with each other to influence the political decisionmakers who had the authority to establish allocation criteria or who had discretionary authority to allocate supplies on their own.

Thus priority users—farmers, providers of mass transportation, providers of emergency services, those involved in telecommunications, and the defense department—were given the right to obtain 100 percent of their "requirements," or in other words, as much as they wanted. As the shortages increased between January and June, the percentage of supplies allocated to priority users by twenty-four refiners responding to a Department of Justice questionnaire increased from 2.76 percent to 3.67 percent.[33] This means that of the 300,000-barrel-per-day increase in gasoline supplied between January and June (which actually was a much smaller increase than normal), about 70,000 barrels went to priority users. Had their percentage of supply not fallen, nonpriority users would have obtained an additional 60,000 barrels per day.

In addition to priority users, others had special status under the allocation rules. Anyone who could demonstrate purchases from a particular refiner, marketer, or jobber during the base period could claim an allocation. Wholesale or end-use bulk purchasers were entitled to 100 percent of their base-period purchases. Other customers on allocation (who in turn were largely the suppliers of ordinary motorists) shared whatever was left over. A refiner would calculate the ratio of current leftover supply to the amount supplied in the corresponding month in the base period. Customers who did not have a right to claim a specific amount from the refiner received a percentage of their base-period purchases equal to this ratio.

Ordinary motorists were not eligible to have supplies directly allocated to them and had to rely on whatever supplies were left over after privileged users had satisfied their own demands. Since price controls prevented them from competing for the available supply by offers of exchange, motorists had to resort to other methods of competition for gasoline—notably competition by willingness to wait in line.

33. U.S. Department of Justice, *Gasoline Shortage*, p. 158.

As was the case in 1974, state set-aside programs assigned each state government 3 percent of all supplies in the state for allocation. In April 1979 this percentage was increased to 5 percent. At first these allocations were mostly awarded to politically favored users, thereby restricting the availability of supplies to motorists even more severely. But after shortages became serious, political pressure forced state governments to allocate most of their set-asides to service stations providing supplies to ordinary motorists. Then, although the set-asides were no longer taking supplies away from motorists, they served to clog the distribution system and delayed supplies from reaching final consumers.[34]

Several changes DOE instituted while the allocations were in effect further aggravated the shortages. The base period for allocations was January 1, 1972, to December 31, 1972. Because of fears that the changes in the geographic pattern of demand since 1972 would lead to shortages if supplies were allocated according to the 1972 pattern of demand, DOE instituted a new base period: July 1, 1977, to June 30, 1978. Numerous objections to this base period led DOE to choose yet another base period: November 1, 1977, to October 31, 1978. Hoping to minimize the effects of demand changes since the base period, DOE also adopted a provision allowing customers whose sales volumes had increased by more than a stipulated amount since the base period to obtain increased allocations. This provision was supposed to allocate additional supplies to customers in areas where demand was increasing most rapidly. Since, however, the months immediately following the base period were winter months, those parts of the country in which driving is relatively heavy in winter—areas with temperate climates and winter resorts—received additional allocations even though the increased driving was attributable to seasonal factors rather than to growth.[35]

Each refiner, moreover, calculated its allocations independently. The allocations depended on a refiner's supply situation and its obligations to supply priority and other privileged users. Thus areas of the country served by refiners whose supplies had been relatively unaffected by the disruptions following the Iranian revolution or which had relatively few supply obligations experienced fewer shortages and dislocations than areas of the country supplied by refiners that

34. U.S. Department of Justice, *Gasoline Shortage*, p. 158 and Lane, *Mandatory Petroleum Regulations*, pp. 82-84.

35. Lane, *Mandatory Petroleum Regulations*, pp. 80-81.

had suffered substantial losses of supply owing to the Iranian upheaval.[36]

Repeating yet another mistake committed in 1973–74, DOE officials encouraged an excessive buildup of stocks during the shortage period. Thus on February 28, DOE urged refiners, "to keep stocks high enough to meet expected demand during the 1979 summer driving season, even if it is necessary to restrict somewhat the amount of surplus [sic] gasoline that is made available to purchasers currently."[37] Between April and July, gasoline stocks increased by 6 million barrels, and at the end of July gasoline stocks were 25 million barrels greater than at the end of July 1978.[38]

Although undoubtedly some would attribute such inventory accumulation to a conspiracy by the oil companies to withhold supplies, the buildup was the result of the incentives created by price controls and allocations, not of any conspiracy. When supplies are uncertain, an incentive exists to increase inventories as a precautionary measure. Should an interruption in supplies actually occur in a market free of price and allocation controls, prices immediately rise. The rising prices reduce consumption and induce those holding inventories in anticipation of a price increase to cash in by drawing down their stocks. Such accumulation during times of relative abundance for use in a period of relative scarcity is precisely what we want to happen. Under controls, however, or when controls are anticipated in case of a disruption, the speculative incentive to hold inventories for the purpose of selling at a profit after an increase in price is eliminated. Some precautionary incentive to hold inventories may persist, but if allocations transfer supplies from those with larger to those with smaller inventories, even this incentive is absent. After any disruption in supplies, since immediate price increases are not possible while gradual price increases can be anticipated, firms have an incentive to build up their stocks instead of drawing them down. Thus, under controls, inventories fluctuate in precisely the wrong way: They fall in times of plenty and rise in times of dearth.

To ensure adequate heating supplies for the following winter, DOE also insisted that refiners build up their distillate stocks to 240 million barrels by October 1979. This figure was based on the assumption that neither government conservation programs nor price in-

36. U.S. Department of Justice, *Gasoline Shortage*, pp. 159–60.
37. 44 Fed. Reg. 11,203 (1979).
38. *Monthly Energy Review*, December 1979, p. 36.

creases would reduce distillate consumption. Even in May, when gasoline shortages were becoming serious, DOE continued to threaten mandatory yield orders if refiners did not produce more distillate. Only in June and July, after gasoline shortages became widespread, did DOE relent and allow increased gasoline production. But in August and September, DOE again pressured refiners to increase distillate yields. What is especially ironic is that although the DOE succeeded in forcing refiners to meet its 240 million barrel target, refiners did so largely by keeping inventories in primary storage at refineries and terminals, where primary stocks are counted, instead of shipping stocks to distributors closer to final consumers. Thus had the fall been unusually cold, shortages of heating oil would still have occurred precisely *because* refiners kept their inventories where DOE could count them rather than where consumers could have had access to them.[39]

CONCLUSION

Though long and somewhat depressing, the story of price controls on and allocations of gasoline and refined products is exceptionally instructive. Unfortunately, its lessons have been ignored or resisted by many of those most in need of them.

The original official explanation for the shortages during 1973–74 was that they were caused by the Arab oil embargo. Although this explanation was plainly contradicted by basic economic theory, it has nevertheless since continued to be widely accepted. So strong has its attraction been that it has very largely shaped American foreign policy toward the Arab world in general and Saudi Arabia in particular. This explanation has been responsible for the notion that Saudi Arabia had the power to inflict gasoline shortages and other punishments on the United States if the U.S. government were insufficiently partial to and solicitous of Saudi interests and wishes. In reality, of course, the Saudis had and have no such power. Only foolish policies cause gasoline shortages. It is true that the Saudis can affect the world price of oil, but there is no reason to believe that the Saudis would charge less than the profit-maximizing price just because the United States adopted policies the Saudis wanted.

39. Lane, *Mandatory Petroleum Regulations*, pp. 76–77.

A second equally erroneous explanation of the shortages has also exercised its influence over public policy. This explanation was that shortages had been the result of an oil industry conspiracy. Charges of conspiracy helped to create a political environment in which it was virtually impossible to terminate price controls on crude oil and refined products. The controls ensured that shortages would be a continuing threat and that, contrary to public policy objectives based on the fear of imported oil, the dependence of the United States on imported oil would be increased.

While hysteria over a meaningless embargo and a nonexistent conspiracy gripped the nation, the simple fact that the shortages were caused by price controls and exacerbated by the political-bureaucratic allocations to which the controls gave rise was totally lost sight of. Further bitter experience and the passage of time have perhaps increased the recognition of this fact somewhat, but it remains as obscure as ever to many.

Yet for those favored by the allocations, the controls provided an escape from the unpleasant necessity of competing by offers of exchange with all other potential users for the available supplies. Those skillful at competing in the political marketplace were able to obtain supplies without having to weigh the value of their own uses for crude oil or products against the values of all other competing uses. Competition for supplies was not eliminated; it was merely shifted from one arena (the market), in which demands are easily registered and alternative costs are thus made explicit, to another arena (the political process), in which only some demands can be registered and alternative costs must therefore remain obscure.

It was thus no accident that motorists bore such a heavy burden during the shortages. Diffuse and unorganized as a group, motorists exercise virtually no political influence, and so it was not likely that the political process would ensure that supplies of gasoline to them would remain adequate. Rather it was smaller more cohesive groups such as farmers that used the political process successfully under controls and allocations to ensure that their access to gasoline remained virtually unrestricted.

Only when the costs of the shortages to motorists became intolerable did the political process begin to respond to their interests. Even then the steps taken to increase supplies to motorists were relatively insignificant. The allocation problem was simply too complex and sophisticated for the political process not to create severe disruptions and dislocations.

It was once said of wage-and-price controls that they are a disaster looking for a country to happen in. The truth of this remark could hardly have been more conclusively demonstrated than by the experience of price controls on refined products. Controls on the prices of crude oil were equally disastrous. After reviewing some historical background, I shall explain in the next two chapters how controls on crude oil prices worked.

Chapter 7

A HISTORY OF PRICE CONTROLS ON CRUDE OIL AND THE ENTITLEMENTS PROGRAM

INTRODUCTION

The interests individuals and groups seek to advance through government intervention in the market are generally so narrow that the full consequences of that intervention are only rarely comprehended by those seeking it. While failure to understand the wider implications of pursuing one's own interests is as much a characteristic of competition by offers of exchange (COE) as it is of political competition (PC), the constraints imposed on COE by a system of private property rights generally ensure that these outcomes reflect most of the information needed to make intelligent decisions about resource allocation. It was precisely this insight that Adam Smith sought to convey with his famous invisible-hand simile. In contrast, the outcomes generated by PC, which operates under constraints far different than those under which COE operates, carry with them no such presumption.

Rarely, if ever, has a regulatory program been undertaken with so little understanding of, and so much confusion about, its effects than in the case of price controls on crude oil and the entitlements program (EP) that accompanied it. Many supported the controls only because they misunderstood its effects, but there were also those who supported controls because, insofar as their own self-interest was concerned, they understood the effects quite well indeed. The ability of these groups to manipulate the political process to advance their own interests is the particular issue I want to address in discussing how price controls came to be imposed on crude oil and maintained for almost a decade. But first, I must discuss the regulatory programs that preceded the crude oil controls, specifically, demand prorationing and the oil import quota systems.

DEMAND PRORATIONING

Early in this century, experts were already warning that the United States was on the verge of an energy crisis because its known reserves would soon be exhausted.[1] However, oil discoveries in Texas, Oklahoma, and Louisiana during the twenties allowed production to expand rapidly to satisfy the enormous growth in demand during that decade without depleting the country's oil reserves. And indeed, the discovery in 1930 of the massive east Texas oil fields ushered in a period of chronic overproduction which, combined with falling demand owing to the Great Depression, led to rapidly falling prices.[2]

Overproduction had long been a common complaint of the oil industry, and under the legal doctrine (the "rule of capture") to which petroleum extraction was subject, overproduction—in a certain sense—was a legitimate problem, not just a slogan invoked by producers seeking higher prices. Overproduction arises because a given oil reservoir normally lies beneath the property of several different owners or lessees. The rule of capture gives each owner or lessee an unlimited right to extract from the common reservoir through wells on the owner's property or leasehold.[3] I have already explained in Chapter 2 why, even if there were no technological repercussions from excessive extraction and drilling, it would be inefficient to treat a reservoir of oil as common property. In fact the inefficiency here is even more serious than the earlier discussion would suggest because extraction reduces the pressure in the reser-

1. See Simon, *Ultimate Resource*, p. 95. Milton Friedman once remarked during a radio interview in the seventies that when he was growing up in the twenties he was taught that there were only enough oil reserves to last for another fifteen years and that now we were down to thirty years.

2. See McDonald, *Petroleum Conservation*, pp. 34–42 and Blair, *Control of Oil*, pp. 160–61.

3. McDonald, *Petroleum Conservation*, pp. 31–32.

voir and increases the cost of extracting the remaining oil. Within certain limits (which define the "maximum efficient rate of production"), this effect is negligible.[4] But beyond these limits increasing the rate of extraction from, and increasing the number of wells in, a given reservoir eventually causes a substantial loss of pressure that eventually increases the cost of extracting the remaining oil to prohibitive levels.

One solution to the problem is to consolidate ownership interests in the reservoir so that current drilling and extraction decisions may be made in light of their repercussions on future extraction from the reservoir. Such consolidation of interests in a reservoir is known as "unitization." Unfortunately, when many separate interests possess drilling rights in a given reservoir, it may be very difficult to secure the unanimous consent of all interests to unitize the reservoir. Instead, the reservoir is likely to remain inefficiently exploited as a common property resource. A practical method of facilitating unitization is to allow a substantial majority of interests in a reservoir to consolidate all interests in the reservoir even if some withhold their consent. Over the years most states have passed legislation allowing unitization with the consent of from 60 to 85 percent of the property interests in the reservoir.[5] But no such legislation was passed until 1945,[6] and Texas still allows only voluntary (i.e., unanimous) unitization.[7]

Other conservation measures have also been enacted. Some (e.g., minimum spacing requirements that limit the proximity of wells to each other or to another property interest within the same reservoir) were pretty clearly directed at limiting waste arising from the treatment of a reservoir as a common property resource. But another measure, demand prorationing—ostensibly instituted to prevent too rapid extraction from commonly held reservoirs—was easily transformed into a device for limiting output in order to raise prices.

Under demand prorationing, which was adopted in most of the oil-producing states, state regulators had authority to determine the output of almost all the wells within their states. In Texas this authority was exercised by the Texas Railroad Commission. The Commis-

4. McDonald, *Petroleum Conservation*, pp. 18–21.
5. McDonald, *Petroleum Conservation*, p. 51.
6. McDonald, *Petroleum Conservation*, p. 217.
7. McDonald, *Petroleum Conservation*, pp. 211–13.

sion would collect demand projections from the Bureau of Mines as well as the planned purchases ("nominations") of the principal oil buyers within the state. Using these and other data, the Commission would, toward the end of each month, estimate statewide demand in the upcoming month and compare it to the state's productive capacity ("basic allowable"). Normally the amount demanded was calculated at a price such that it would be less than the basic allowable for the state. To avoid overproduction, the Commission would calculate a "market-demand factor," representing the estimated ratio of market demand to the state's productive capacity. The productive capacity of each well (aside from those that were exempt for some reason) was multiplied by the market-demand factor. The result was the allowable output of each well in the state for the upcoming month. Thus each well had a pro rata share, based on the well's productive capacity, of the state's total demand.[8]

Because Texas was by far the nation's dominant oil-producing state, accounting for 43.2 percent of the nation's crude oil output and 54.8 percent of its crude oil reserves between 1949 and 1951,[9] the Texas Railroad Commission could effectively determine domestic crude oil prices. Before the emergence in the late forties of the Persian Gulf and Venezuela as major international oil suppliers, the Commission in effect operated on behalf of Texas oil producers as the first international oil cartel.

The Texas cartel was undermined by the entry of supplies from the Persian Gulf into world oil markets. These supplies forced down the world price below United States levels, with the result that the United States was soon transformed from a net exporter to a net importer of oil. The domestic price, of course, came under pressure from supplies imported at the world price. Regulators would have had to impose increasingly severe limitations on domestic output to maintain domestic prices. By 1954 the situation had become so serious that domestic oil producers began to call for limits on imports of crude oil into the United States.

8. McDonald, *Petroleum Conservation*, pp. 48-49. The Texas Railroad Commission still exerts formal control over production levels in Texas, but it can no longer exert any control over domestic prices. For the past decade allowables have been set at 100 percent of capacity.

9. McDonald, *Petroleum Conservation*, p. 43.

THE IMPORT QUOTA SYSTEMS

Since the early fifties, whether to limit oil imports and, if so, how have been among the leading issues facing U.S. policymakers. Although for almost two decades the primary effect of limiting oil imports was to enrich domestic oil producers, that was not the policy's ostensible purpose.

The Rationale for Limiting Imports

According to four Presidents (Eisenhower, Kennedy, Johnson, and Nixon) limiting oil imports was vital to the national security of the United States. Their reasoning was that allowing unrestricted imports to increase would leave the country vulnerable to an undue risk that the oil imports would be cut off in the event of war. Furthermore, falling domestic prices would reduce domestic production and exploration and would, thus, lead to increasing U.S. dependence on oil imports over time.

Even granting the good faith of those advocating the national security argument for limiting oil imports – the mere mention of which, it has been said, would elicit uproarious laughter at the Petroleum Club – one may still question the policy of limiting imports by quotas. In approaching the issue one has to ask, first of all, if any limit on oil imports served the national security interests of the United States. Despite the pronouncements of four Presidents, the answer is not obviously yes. And even if it were, the case for quotas would still not have been made for there may well have been preferable alternative methods of limiting imports.

To answer the first question would require far more attention than I can devote to it here,[10] so I shall simply make one or two general observations that are sometimes overlooked in discussions of the issue. While there were, to be sure, legitimate security issues raised by the prospect of an increasing dependence on foreign oil, limiting imports was not necessarily the only way of addressing those concerns. Stockpiling oil or shutting in capacity for use only in case of

10. The issue is dealt with at length in Bohi and Russell, *Limiting Oil Imports*, especially Chapter 9.

an emergency while continuing to import oil without restriction were alternatives.[11] Also, limiting oil imports implied a decision to drain the United States first and, hence, ultimately to increase the dependence of the United States on foreign sources of oil.

It could be argued that limiting imports and raising prices increased the incentive to find domestic reserves and thus increased national security in the long run. According to this argument, raising the current price for oil would encourage exploration for reserves to be available in case of emergency. The problem with this argument is the implicit assumption that potential explorers for oil lack the appropriate incentive to search for domestic reserves that could become extremely valuable were access to foreign supplies to be interrupted. That assumption would only be true if the probability of interruption were systematically underestimated by the market or if those who discover new oil reserves could not establish secure property rights over their discoveries. Otherwise the anticipation of earning high profits from oil that would become very valuable in case of interruption would be expected to induce the appropriate amount of exploration for and voluntary stockpiling of oil. Unless the government has inside information that the probability of war (or interruption of supplies in case of war) is higher than the public expects, I see no reason why the market would underestimate the likelihood of an interruption of supplies.

It is likely, however, that oil explorers and potential stockpilers do not believe that property rights over oil reserves they might discover would be secure either because of the common resource problem or, perhaps more importantly, because they fear reserves would be expropriated in an emergency. Expropriation could be carried out directly, but it is likely to be carried out indirectly either through price controls or taxes.

In view of such fears by potential oil explorers, which it would be difficult to dispel no matter what the current intentions of any government, there might be some grounds for subsidizing oil exploration.[12] Limiting oil imports to raise domestic prices is only one of many conceivable ways of doing this, nor is it obviously superior to the alternatives.[13]

11. Bohi and Russell, *Limiting Oil Imports*, pp. 317–22.

12. See Thompson, "National Defense Argument," for an argument rationalizing subsidies to certain domestic industries based on underinvestment because of the expectation of wartime price controls.

13. Bohi and Russell, *Limiting Oil Imports*, pp. 322–28.

Given the decision to limit imports, however, quotas were certainly not the most efficient method to use. The national security objective of limiting oil imports could just as well have been satisfied by an appropriate tariff on imports of crude oil and refined products. The volume of imports would have been reduced; producers would have benefited from a higher price; and the Treasury would have collected the difference between the foreign price and the landed price of every barrel of imported oil.

Even under an import quota system, rights to import could have been auctioned off by the Treasury, thereby collecting (as it would have under a tariff) the difference between the foreign and domestic prices of each barrel of imported oil.[14] Instead, the quota tickets were awarded gratis to various privileged recipients who captured the revenue that could just as well have been collected by the Treasury. Although ostensibly imposed to protect national security, the import quota program–particularly the disposition of the rights to import–inevitably became ever more deeply enmeshed in the tangle of special interests competing in the political process.

The response of the political process to the quota system can be described succinctly. If you leave money lying around in public, someone will try to pick it up. And the longer you leave it there, the greater will be the number of those trying to pick it up. Instead of restricting the subsidy inherent in the limitation of imports to producers who could arguably have contributed to the national security objectives of the program, the political process allowed first refiners and later various other groups to share in the wealth transfer inherent in the program. Let me now briefly describe the historical development of the oil import quotas.[15]

14. The difference between a tariff and a quota system under which the quotas are auctioned off is that under the former a domestic price is set (the world price plus the tariff), and the volume of imports required to satisfy domestic demand at the price is determined implicitly; whereas under a quota system the volume of imports is set, and the domestic price is determined implicitly. If market conditions are known in advance, the domestic price and domestic consumption would be the same under either system. However, if market conditions are uncertain, the tariff implies that the uncertainty will be reflected in variability in the volume of imports, while under the quota the uncertainty will be reflected in the variability of the domestic price.

15. For more complete accounts of the oil import quotas see Bohi and Russell, *Limiting Oil Imports*, and Dam, "Import Quotas." A narrower, more technical but very useful study of the effects of the quotas in Barzel and Hall, *Political Economy*.

The Voluntary Programs

In 1954 President Eisenhower established the Cabinet Advisory Committee on Energy Supplies and Resource Policy to study the problem of growing dependence on imported oil and to make recommendations for safeguarding U.S. security interests affected by oil imports. The committee report urged that oil imports, as a percentage of domestic production, not be permitted to exceed the 1954 level. The report also recommended that import restriction be achieved through the voluntary restraint of importers. Following this recommendation, the Office of Defense Mobilization asked oil importers to reduce their imports from outside the Western Hemisphere by 7 percent during the second half of 1955. Nevertheless, oil imports continued to increase despite protestations by importers that they were exercising restraint.[16]

Alarmed by the growing market share of imports, domestic oil producers intensified their campaign for mandatory controls on oil imports. In 1957 the administration responded to calls for tighter controls on imports by adopting a formal, though still "voluntary," system of import control. The program was voluntary inasmuch as there were no legal sanctions against those exceeding their quotas. The only deterrent to noncompliance was the understanding that if the voluntary program proved ineffective, a mandatory system would replace it.

In administering the system, the Treasury department divided the United States into five geographic districts. District V, consisting of the Pacific Coast states plus Nevada and Arizona, was technically exempt from the quota system. But districts I–IV, encompassing the rest of the (then) forty-eight states, were subject to formal import limitations.

The voluntary program defined three classes of importers: established importers, small importers, and newcomers. The established importers (Atlantic, Gulf, Sinclair, Mobil, Standard of California, Standard of New Jersey, and Texaco) were "requested" to import 10 percent less crude oil than they had been importing on average between 1954 and 1956. Small importers were permitted to import as much as they had planned in the last half of 1957 provided that

16. Bohi and Russell, *Limiting Oil Imports*, pp. 26–30.

they did not increase imports between 1956 and 1957 by more than 12,000 barrels per day. Newcomers were supposed to submit requests to the Treasury for permission to begin importing crude oil. Growth rate of imports by newcomers was made contingent on future expansion of the domestic market so that imports by newcomers would not encroach on the imports of established importers.[17]

Despite formally assigning quotas to refiners, the second voluntary program was no more successful than its predecessor in limiting imports. It was undermined by the outright noncompliance of some importers; by the incentive, owing to the differential between the prices of imported and domestic crude oil, for new importers to enter the market; by the lack of restraints on importing refined products; and by the potential antitrust violations implicit in the voluntary program that led the Justice Department to recommend its replacement with a mandatory program.[18]

The success of the voluntary program was, after all, inconceivable unless the oil industry abided by a collusive agreement to raise prices. The voluntary quotas asked oil importers to forego the immediate tangible profits to be earned by importing low-cost oil and selling the products refined from that oil at prices reflecting the higher cost of domestic crude oil. In return for such restraint, oil importers were promised longer lasting profits on the difference between low foreign and high domestic prices and any patriotic satisfaction that might be derived from acting in a way that ostensibly increased national security. Had the capacity to ensure compliance with the voluntary program rested with a single firm or with a group of colluding firms, such a firm or group might have continued to comply with the voluntary program. As it was, the large number of firms and the relative ease of entry made it obvious to every firm that the voluntary program would succeed or fail quite independently of the extent of its own compliance with the program. Thus, competition made the ultimate breakdown of the voluntary program inevitable.

The Mandatory Program

The breakdown of the voluntary program led to the imposition in March 1959 of the Mandatory Oil Import Program (MOIP). To pre-

17. Dam, "Import Quotas," pp. 6-8.
18. Dam, "Import Quotas," p. 14.

vent importers from evading the restrictions on importing crude oil by importing refined products, MOIP limited the importation of products as well of crude oil. MOIP retained the five-district division of the United States for administering the system, and district V continued to be subject to different rules than the other four districts. Total imports of crude oil and refined products into districts I–IV were not to exceed 9 percent of domestic demand. This objective, which was never met, would have required an absolute decline in petroleum imports to the United States.[19]

MOIP made it illegal to import crude oil or refined products without an import quota ticket issued by the Treasury. The number of quota tickets determined the volume of imports and the level of domestic prices for both crude oil and refined products. Increasing the quota reduced domestic prices for crude oil and refined products. This is an important relationship because it was responsible for one of the underlying tensions in the effort to limit imports. While domestic producers favored reducing the quota in order to raise domestic prices, refiners had reason to favor relaxing the quota and reducing prices since reducing product prices would increase the demand for refining services to convert crude oil into products.

This does not mean, of course, that refiners necessarily were harmed by quotas. Insofar as quotas enabled them to acquire cheap foreign oil to convert into products they could sell at high domestic prices, the quotas could be a source of profits for refiners. It was the distribution of quota tickets that determined whether refiners, on balance, gained or lost from quotas. How the quota tickets were distributed was therefore as crucial to the viability of the program as the amount of imports allowed. One way of distributing the tickets would have been to auction them off to the highest bidders. This would have captured what we may call the scarcity value of the quota for the Treasury and, hence, for taxpayers in general.

Distributing the tickets was more complicated than it sounds because to divide up the quotas among a diverse group of refiners with sharply divergent interests without turning any significant number of them against the program was itself no simple task. For the program to be viable, three groups of refiners had to be won over, or at least placated: established importers, large refiners that had not previously imported large amounts of crude oil, and small refiners.

19. Dam, "Import Quotas," p. 15.

The established importers wanted the allocation of quotas to reflect the amounts that refiners were already importing. Otherwise, established importers would be greatly disadvantaged by the loss of cheap imports on which they had counted and in the expectation of continued access to which they had invested. But large refiners that had not previously imported substantial amounts of oil opposed historically based allocations that would have granted established importers a nearly exclusive privilege to import oil. They argued, instead, for awarding quotas in proportion to the refining capacity of each refiner. For their part, small refiners maintained that their small size entitled them to special treatment (so as to promote competition within the industry), and so they demanded a share of quota tickets greater than their share of refining capacity.

To maintain support for the quota system from each substantial interest, the administration fashioned an allocation system that in some measure satisfied the historical claims of the established importers, the proportional claims of the large newcomers, and the claims of the small refiners for special treatment. The historical claims were satisfied by guaranteeing that no firm would receive quotas entitling it to import less than 80 percent of what it had been permitted to import under the voluntary program. It was envisioned, however, that this preference for established importers would gradually be phased out over some adjustment period of unspecified length. But though diminished, the historical element continued in the allocation of quotas until the MOIP was abolished in 1973.[20]

The two other main groups of refiners were reconciled to the system by providing that, apart from historical quotas, quota tickets would be allocated to refiners based on their refining capacity—but the distribution was not strictly proportional. A so-called sliding-scale awarded proportionately more import quotas to small refiners than to large ones.

For some refiners the allocation of import quota tickets would have been of little value had they been required to refine the oil they imported under MOIP in their own facilities. An inland refiner that had to transport imported oil to its own refinery would have dissipated a substantial portion of the value of its import quota tickets in transportation costs. To preserve the value of quota tickets for

20. Dam, "Import Quotas," pp. 17-24 and Bohi and Russell, *Limiting Oil Imports*, pp. 114-25.

inland refiners, it was therefore essential that the tickets be in some measure transferable. So, although tickets could not be sold for cash, they could be exchanged for oil.

A typical arrangement was for an inland refiner to import oil for delivery at some coastal port. Instead of transporting the oil to its inland refinery, the importer would deliver the oil to an integrated refiner with a refinery near the port in exchange for an equivalent amount of crude that the integrated refiner could deliver to the inland location from a nearby domestic source. An inland refiner receiving, say, 10,000 tickets a month might import 10,000 barrels of crude oil for $2 per barrel and exchange the imported oil for an equivalent amount of nearby domestic oil, for which it would have had to pay perhaps $3.25 per barrel in a money transaction. In essence, the quota amounted to a $12,500-a-month subsidy to the inland refiner.[21]

Although the method of allocating quota tickets may not have guaranteed that every refiner would be a net beneficiary of MOIP, it did spread the benefits widely enough to ensure that most gained, while few lost, because of it. A cardinal principle for refiners was that the tickets be awarded only to refiners. The importance of the principle was derived not so much from group solidarity as from the fear that the subsidy might be diluted by a rapid increase in the number of recipients. Since the subsidy itself was not large enough to justify the cost of building a new refinery, observance of the principle would prevent any substantial increase in the number of recipients.[22]

From its inception, MOIP was surrounded by controversy. Despite the adverse effects MOIP had on consumers, most of the controversy concerned intraindustry competition for shares in the subsidies generated by the program and the industry's efforts to exclude outsiders from sharing in the subsidies. Nevertheless, competition for benefits generated by MOIP ensured that the quotas would inevitably be shared by an increasing number of recipients.[23]

Competition for quota tickets also elicited an increase in the number of tickets issued, which gave imports a growing share in the

21. The value of a ticket was widely thought to be worth roughly $1.25 during most of the life of the MOIP. See Barzel and Hall, *Political Economy*, p. 55.

22. See Stigler, *The Citizen*, p. 115 for a statement of the general principle.

23. See Barzel and Hall, *Political Economy*, pp. 31-50 and Dam, "Import Quotas," pp. 44-52.

domestic market. Consequently, the real domestic price of crude oil declined between 1960 and 1970.[24] This was not exactly the outcome that oil producers had expected MOIP to result in. Responding to growing dissatisfaction with MOIP, President Nixon appointed a Cabinet Task Force on Oil Import Control, headed by Labor Secretary George Shultz, to study the question of oil imports.[25] But when early in 1970 the task force recommended that MOIP be replaced by a simple tariff on imported oil, Mr. Nixon rejected the proposal and pledged to retain the quotas. Since the nonproducer beneficiaries of MOIP would not have countenanced import limitations without continuing to receive such compensation as MOIP still provided them, Mr. Nixon and the domestic oil producers were unwilling to replace MOIP with a tariff on imported oil. Despite almost everyone's dissatisfaction with the implementation of MOIP and despite the theoretical superiority of a tariff as a means of limiting imports, no alternative to MOIP seemed to have the support of a viable coalition of interests. What is perhaps most puzzling is why Mr. Nixon appointed the task force in the first place and what recommendations he could have been expecting it to make.

The Collapse of the Mandatory System

Whatever Mr. Nixon's expectations about MOIP, they were soon overwhelmed by events. Domestic crude oil production began to decline after 1970, while domestic demand increased rapidly.[26] At the same time, the excess production capacity that had long been maintained by demand prorationing was rapidly being used up so that by 1973 monthly allowables determined by the Texas Railroad Commission reached 100 percent of capacity.[27] If severe shortages or sharply higher domestic prices for refined products—both politically unacceptable—were to be avoided, imports had to be allowed to increase rapidly.

Profound changes were also occurring in the world oil market. The Organization of Petroleum Exporting Countries (OPEC) had been formed in 1963 to bargain with the oil companies for more favorable

24. Bohi and Russell, *Limiting Oil Imports*, p. 211.
25. Cabinet Task Force on Oil Import Control, *Oil Import.*
26. Bohi and Russell, *Limiting Oil Imports*, pp. 22–23.
27. Owens, "Petroleum Price Controls," p. 1305.

terms. Initially it had met with indifferent success at best, but late in 1970 Libya achieved a breakthrough by forcing the oil companies operating there to agree to an increased price for Libyan oil.[28] Thereafter, OPEC moved quickly to take control of world oil markets. From less than $2 per barrel in late 1970, the world price of oil rose to over $10 per barrel by the beginning of 1974. As early as the spring of 1973, the landed price of imported oil matched that of domestically produced oil.

Once world oil prices rose to U.S. levels, import quota tickets lost their value to domestic refiners, and quota ticket recipients were deprived of the subsidy they had enjoyed for more than a decade. As I explained in Chapter 5, generally rising prices for crude oil, the loss of the subsidy from import quota tickets, and the shortages arising from price controls (particularly special rule number one) on crude oil and refined products combined to throw the domestic refining industry into chaos. In an attempt to make supplies more accessible, Mr. Nixon terminated MOIP in April 1973. In its place he imposed a 10.5¢-per-barrel license fee on imported oil. The fee was gradually increased to 21¢ per barrel by 1975. Recipients of quota tickets were exempted from payment of the fee on imports up to their 1973 quota allocations.[29] But even without the quota, price controls kept petroleum markets in turmoil throughout 1973.

The loss of their subsidy from the import quota tickets was a severe hardship for many refiners, particularly the small inland ones that had received disproportionately large subsidies under MOIP. Without cheap imported oil to exchange, many inland refiners could not obtain the amounts of crude oil from domestic sources that they had previously been using. Having already learned how to use the political process to advance their economic interests, domestic refiners lobbied Congress for an allocation system that would guarantee them a share of the domestic supply of crude oil. By November 1973, during the Arab oil embargo, the pressure of domestic refiners combined with a general sense of panic to ensure passage of the Emergency Petroleum Allocation Act (EPAA). Moreover, the combination of rising world oil prices and controls on domestic oil prices were creating conditions that would allow a renewal of the subsidies refiners had apparently lost with the demise of MOIP.

28. See Adelman, *World Petroleum Market*, pp. 250–56, and Sampson, *The Seven Sisters*, pp. 248–73.

29. Bohi and Russell, *Limiting Oil Imports*, pp. 230–34.

PRICE CONTROLS ON CRUDE OIL

When controls were first imposed on domestic oil prices in August 1971, domestic oil prices were still substantially above world prices. There was therefore little upward pressure on crude oil prices domestically, and controls had little impact on market conditions. In addition most domestic oil producers were exempted in May 1972 from price controls under the Small Business Exemption.[30] Only when, early in 1973, world oil prices reached and then surpassed domestic levels did the price controls begin to have much impact on crude oil markets.

While oil produced by the larger producers was subject to controls, a very substantial part of domestic oil production was undertaken by uncontrolled producers. The price of uncontrolled domestic oil was thus rising rapidly along with the world price during the spring of 1973. A further increase was imminent when, in June 1973, President Nixon imposed a second freeze on wages and prices. During the freeze, the Cost of Living Council (CLC) concluded that controlling the price of oil sold by major oil companies while allowing the price of oil sold by smaller firms to respond to market forces (as special rule number one had required) was an unworkable policy. Neither willing nor able to abandon controls entirely, CLC had no choice except to extend controls to the entire industry.

Even while adopting a more comprehensive set of price controls under Phase IV, however, CLC continued to control some oil prices but not others. CLC now distinguished between old oil (controlled) and new oil (uncontrolled). As of September 1973 oil extracted during any month from a producing property up to the amount extracted from the property in the corresponding month in 1972 was defined as old oil. Monthly production from a property exceeding production during the corresponding month in 1972 was defined as new oil.[31] All oil produced from properties that began production after 1972 was counted as new oil. Furthermore, every barrel of new oil produced from an old property entitled the producer to release from controls one barrel of old oil from the same property.

30. Owens, "Petroleum Price Controls," p. 1236.

31. The rule was somewhat more complicated since cumulative production for the year also had to be above cumulative production for 1972 up to the corresponding month. See Owens, "Petroleum Price Controls," p. 1265 and Kalt, p. 12.

CLC set the ceiling price for old oil equal to the price that oil from the same property had sold for on May 15, 1973, plus 35¢ per barrel. In December 1973 the ceiling price for old oil was increased by $1 per barrel. This raised the average ceiling price for oil to $5.03 per barrel (not $5.25 per barrel as it was erroneously believed to be at the time).[32]

With controlled and uncontrolled markets for physically indistinguishable commodities, there were powerful incentives to evade the controls. One technique was for a producer to require customers to purchase new oil for more than the market price as a condition for purchasing old oil so that the average price for the entire sale would equal the market price for uncontrolled oil.[33]

As markets became increasingly chaotic, owing to restrictions on COE, PC became correspondingly more intense. Nonintegrated refiners and marketers began lobbying for allocation of crude oil and products. At first the pressures for allocation were met with voluntary programs instituted by a number of major oil companies. These programs gave established customers some share of the available supplies based on their share of supplies in an earlier period. The situation grew still worse following the outbreak of the Yom Kippur War and the cutback in production by the Arab oil exporting countries. The ensuing disruption led quickly to the passage in November 1973 of EPAA, which imposed mandatory allocations on the oil industry.[34]

In an amendment to the Alaska Pipeline Bill, Congress lifted price ceilings from stripper wells (i.e., wells that produce fewer than ten barrels per day). The exemption was justified on the grounds that without it many such wells would have been shut down because of the high cost of extracting oil from them. Of course the fact that those wells were mostly owned by independent oil producers (who had been among the chief beneficiaries of MOIP and who were not stigmatized as were the major oil companies) did not harm the chances of the exemption's approval by Congress.

EPAA required that all crude oil and refined products be "totally allocated for use by ultimate consumers within the United States..." in accordance with the objectives specified by the act and at prices

32. Kalt, *Oil Price Regulations*, p. 12 and Bohi and Russell, *Limiting Oil Imports*, p. 222, footnote 76.

33. See Owens, "Petroleum Price Controls," pp. 1306-07.

34. Lane, *Mandatory Petroleum Regulations*, pp. 25-27.

set by regulations.[35] I have already described the allocation of refined products, so I shall limit my comments here to the allocation of crude oil. The Federal Energy Office (FEO), which was supposed to draft allocation guidelines for the oil industry, required that all crude oil producers continue selling at controlled prices to their customers as of December 1, 1972. Every customer was to be given the right to purchase the same percentage of a producer's total crude output that it had purchased in December 1972. By partially vesting property rights to old oil with specific purchasers, FEO eliminated the incentive for purchasers to agree to buy old oil under tie-in arrangements with new oil such as those I mentioned above. The guidelines also eliminated the incentive for crude oil producers to sell their properties to refiners as a way of evading controls on old oil prices since they prevented the refiner from using the old oil itself and required the refiner to continue selling the old oil to the producer's December 1972 customers.[36]

Although the controls on old oil prices were supposed to have been gradually relaxed so that old oil prices could rise to world market levels, the dramatic rise in world oil prices in the fall of 1973 increased the stake of those benefiting from the controls in their perpetuation. The gap between the price of old oil and the uncontrolled price implied too great a wealth transfer not to generate intense political competition.

The substantial price differences between categories of crude oil also created huge disparities in earnings among refiners based on the share of old oil in their crude oil input. Owing to the FEO allocation rules, such variations were a matter of historical happenstance. Disadvantaged refiners protested against the injustice of having to pay so much more for crude oil than did their more fortunate competitors. Nor did they protest in vain. Refiners used the political process to compete with each other for the economic rents derived from refining old oil just as they had used it to compete for the rents they had derived from importing oil under MOIP.

The first formal attempt to redistribute the rents from refining old oil that this competition gave rise to was the buy-sell program instituted by FEO in February 1974. Under the program, refiners with more than the average rate of utilization of refining capacity were

35. Public Law Number 93-159 (November 27, 1973), Section 4(d), 87 Stat. 629 (1973) quoted in Lane, *Mandatory Petroleum Regulations*, p. 38.

36. Kalt, *Oil Price Regulations*, pp. 12-13.

required to sell oil to refiners with less than the average rate of utilization of refining capacity. A refiner obliged to sell had to do so at a price equal to the average acquisition cost of all the oil it had obtained from all sources. Since these transactions took place at below-market prices, they involved a wealth transfer from sellers to buyers. Insofar as refiners obliged to sell under the programs had relatively more old oil to refine than those that bought under the program, the buy-sell program did redistribute some of the gains from controls on old oil among refiners.

However, the program did not work out in all respects precisely as it had been intended to. One unforeseen consequence of the program was that it discouraged imports of crude oil because imports either increased the amount of oil a refiner was obliged to sell or reduced the amount it was permitted to buy under the program. It also turned out that the program allowed some major oil companies to be buyers and required a number of small refiners to be sellers of crude oil. That, too, was not exactly what had been anticipated. Given both a general political climate hostile to the major oil companies and the considerable political influence of the small refiners, such a result was not tolerated for long. Thus in May 1974 the program was revised so that only the fifteen largest refiners would be sellers, and none of them could be purchasers under the program.[37]

THE ENTITLEMENTS PROGRAM

The failure of the buy-sell program to satisfactorily redistribute the benefits derived from the controls on old oil prices among refiners did not mean that the cause was abandoned. To understand how the next attempt worked, recall that refiners had been subsidized under MOIP through the distribution of valuable rights to import a limited amount of relatively cheap oil. Unfortunately for refiners, MOIP and the subsidy it made possible were swept away by rising world oil prices in the early seventies.

The price controls on old oil can be viewed as the means by which refiners—having lost their import quota subsidy and having incurred capital losses on refinery equipment because of the rising price of crude oil—sought to transfer back to themselves part of the windfall

37. See Lane, *Mandatory Petroleum Regulations*, pp. 40–43 and Kalt, *Oil Price Regulations*, p. 13.

the rising price was conferring on producers. In giving refiners access to a limited amount of relatively cheap oil, the price controls also made it possible to subsidize refiners by precisely the same technique—the allocation of rights to cheap oil—that MOIP had used earlier. The competition among refiners for access to this limited amount of price-controlled oil was of course similar, if not identical, in character to the competition for import quotas under the MOIP. One would therefore expect to find in the contours of the entitlements program (EP)—the ultimate result of this competition—certain telltale signs of similarity between the forces that shaped it and those that determined the distribution of quotas.

Let me now proceed to give a brief and somewhat simplified description of EP. Without EP, a refiner could only have benefited from price controls and old oil insofar as it could physically refine old oil into products and take advantage of the greater markup on old oil than on new or imported oil. But mandatory allocations had frozen the physical distribution of old oil into its pattern as of December 1, 1972. This meant that refiners processing large amounts of old oil received more of the benefits from price controls than did refiners processing small amounts of old oil. What EP was designed to do was to render the distribution of benefits among refiners independent of the actual physical distribution of the old oil itself.

EP did this by prohibiting a refiner from refining any old oil without having an entitlement for each barrel of old oil refined. Every month FEO (and later its successors FEA and DOE) would distribute just enough entitlements to domestic refiners to refine all the old oil available that month. But since the entitlements were distributed to refiners in proportion to the total amount of oil they refined (with some qualifications to be mentioned later), refiners that used relatively large amounts of old oil in their refineries had to purchase entitlements from refiners that used relatively little old oil in their refineries. Entitlement sales redistributed income from refiners that refined relatively large amounts of old oil to those that refined relatively small amounts of old oil. Had entitlements in fact been distributed strictly in proportion to each refiner's total use of crude oil, the subsidy from refining old oil would have been redistributed among refiners so that it was proportionate to the amount of oil processed by each refiner.

Under EP, just as under MOIP, all refiners were equal, but some were more equal than others. In particular, the small-refiner bias that had been characteristic of MOIP was carried over into EP. Under EP the small-refiner bias, at various times, either allocated extra entitlements to small refiners (those with refining capacity of less than 175,000 barrels per day) or allowed them to refine old oil without entitlements, which meant they could sell their entire allocation of entitlements.[38]

However, just as they had been unable to prevent nonrefiners from sharing in the allocation of import quotas, refiners were unable to prevent nonrefiners from sharing in the distribution of entitlements. Thus one method of reducing heating oil prices in the Northeast, where heating oil prices were a particularly sensitive issue, was to distribute entitlements to importers of residual fuel and heating oil.[39] Because EP tended to reduce the demand for low-quality old oil, a special entitlements subsidy was granted to purchasers of low-quality California crude, which had previously not been marketable at the national ceiling price for old oil.[40] Finally, as a means of financing accumulation of the strategic petroleum reserve, the federal government awarded itself entitlements on the importation of crude oil for storage in the reserve.[41]

FURTHER CHANGES IN CONTROLS ON CRUDE OIL PRICES

The government's authority to control petroleum prices and to allocate supplies under EPAA was originally scheduled to expire in August 1975. Unable to agree on a program for extending controls and unwilling to tolerate the immediate decontrol that would have ensued if EPAA had been allowed to expire, Congress (which wanted to extend controls) and the Ford Administration (which wanted to phase them out over time) agreed on a four-month extension of EPAA until the end of 1975.

After months of intense negotiations, Congress finally passed, in December 1975, an extension of controls—the Emergency Petroleum

38. Lane, *Mandatory Petroleum Regulations*, pp. 131-36 and Kalt, *Oil Price Regulations*, pp. 59-61. Stigler (*The Citizen*, p. 119) conjectures that a general feature of all regulatory programs that protect industry interests is that smaller firms receive proportionately more of the benefits than larger ones.

39. Kalt, *Oil Price Regulations*, pp. 61-62.

40. Kalt, *Oil Price Regulations*, pp. 63-64.

41. Kalt, *Oil Price Regulations*, p. 64.

Conservation Act (EPCA)–which did provide for ultimate decontrol on September 30, 1981. It also granted the President authority to decontrol refined products and authority to place crude oil price controls on standby after May 30, 1979. In several respects, however, EPCA actually tightened and extended price controls on crude oil.

Under EPCA there were three categories of domestic oil: lower tier oil, upper tier oil, and decontrolled oil. Lower tier oil was defined as oil produced in any month from properties not exceeding the lesser of (1) the average monthly amount of old oil produced in 1975 or (2) the average monthly amount of oil produced in 1972. Upper tier oil was defined as oil produced from any pre–1976 properties above the maximum lower tier output and all production from properties that began production after 1975. EPCA also terminated the "released oil" provision of EPAA that had allowed one barrel of old oil to be released from controls for every barrel of new oil produced. Lower tier oil continued to sell at its May 15, 1973, price plus $1.35 per barrel plus inflation and incentive adjustment factors to be determined by the Federal Energy Administration (FEA). Upper tier oil was to sell at its September 30, 1975, price minus $1.32 per barrel plus inflation and incentive adjustment factors to be determined by FEA. Stripper oil was originally treated as upper tier, but in September 1976 it was decontrolled. Production from federal naval petroleum reserves and, starting in September 1978, production attributable to tertiary oil recovery projects were decontrolled. The average price of all domestic oil was not to exceed $7.66 per barrel plus inflation and incentive adjustment factors. Incentive adjustment factors could be allocated between lower and upper tier oil prices at the discretion of FEA, but the incentive adjustment factors could not exceed 3 percent, and the combination of the inflation and incentive adjustment factors was not to exceed 10 percent. In September 1976 EPCA was amended to allow the average domestic price to rise by up to 10 percent per year without regard to the inflation rate or to limits on incentive adjustments.[42]

The change in price controls under EPCA also required certain alterations in EP. Since there were now two categories of oil subject to different price ceilings, the number of entitlements issued to refiners in any month was now to equal the number of barrels of lower

42. Kalt, *Oil Price Regulations*, pp. 15–16.

tier oil refined in that month plus a fraction of the number of barrels of upper tier oil refined in that month.[43]

In June 1979, under provisions of the EPCA, President Carter began the gradual decontrol of domestic crude oil prices. Mr. Carter immediately decontrolled oil produced from onshore properties not producing before 1979 and from offshore properties leased after 1978. Beginning in January 1980 a portion of every property's output of upper tier oil was to be decontrolled each month. The calculation of lower tier production was changed, beginning in January 1980, so that each month less of any property's output was to be counted as lower tier and more as upper tier oil. Mr. Carter also adopted various other measures to lift controls from oil prices, either immediately or over time.[44]

In addition to the gradual decontrol which, under EPCA, he had authority to implement on his own, Mr. Carter also requested that Congress enact a windfall profits tax (WPT) on the increased revenue accruing to sellers of crude oil owing to the decontrol of domestic crude oil prices he was ordering. In effect, WPT took the form of an excise tax on the difference between the price of oil before controls were lifted and the uncontrolled price of that oil. After months of Congressional debate and negotiations, a compromise bill embodying most of Mr. Carter's recommendations for WPT was enacted by the Congress in January 1980.

Shortly after his inauguration as President, Ronald Reagan ordered the complete decontrol of all domestic crude oil and refined product prices—thus ending almost ten years of direct controls over petroleum prices. Although he earlier had campaigned against WPT, as President, Mr. Reagan was unwilling to ask Congress for its repeal. Thus from the standpoint of the producer, the price of crude oil is still controlled, and at least one remnant of the misguided enterprise continues to play a role in the market for crude oil.

43. The fraction equaled the difference between the average prices of decontrolled oil and upper tier oil divided by the difference between the average prices of decontrolled oil and lower tier oil. The price of an entitlement was set equal to the difference between the refiners' average acquisition cost of uncontrolled oil and the average price of lower tier oil.

44. Kalt, *Oil Price Regulations*, pp. 16–17.

CONCLUSION

Government intervention in the market for crude oil predates the imposition of price controls in 1971. While market intervention before 1971 tended to increase the domestic price of crude oil, after 1971 intervention was aimed at reducing the domestic price. This might suggest that before 1971 intervention was carried out in behalf of producer interests but that after 1971 it was carried out in behalf of consumer interests. Such an inference would ignore important elements of continuity in the pre- and post-1971 experience. Although crude oil producers were treated less favorably after 1971 than they had been before, refiners were able to turn the post-1971 controls, designed to reduce the price of crude oil, into a refiner subsidy just as they earlier had succeeded in turning the oil import quotas, designed to raise the price of crude oil, into a refiner subsidy.

What is more, it is highly doubtful that consumers – the supposed beneficiaries of the post-1971 controls – derived any net benefits from the controls. The belief that they did stems largely from an incorrect theory of price determination, and although EP may have had some tendency to reduce consumer prices, it also had countervailing tendencies to increase consumer prices. In the next chapter I shall explain how price controls and EP affected consumer prices and provide some evidence to show that their net effect was, indeed, to raise consumer prices.

Chapter 8

A GUIDE TO PRICE CONTROLS ON CRUDE OIL AND THE ENTITLEMENTS PROGRAM

INTRODUCTION

The recognition that the essential function of market prices is to convey information about the world as it is and to induce people to make often unpleasant adjustments that those conditions call for leads one inevitably to conclude that price controls are a kind of escapism—an attempt to avoid coping with conditions in the world as it is. An increase in the price of any commodity informs us of an excess demand for it. The excess demand might be due to an increase in demand or a reduction in supply, but in either case the higher price informs people that some of the uses they had been making of the commodity are no longer worthwhile and that perhaps they ought to find substitutes for it or ways of economizing on its use. The higher price also informs people that some resources formerly used to produce other commodities can now more profitably be used to produce this one.

Since certain activities that had formerly been undertaken are no longer profitable, some of the adjustments to the new state of the world will undoubtedly be painful. Those for whom the adjustments are most painful, unaware or unheedful of the underlying reality reflected in the higher price, naturally seek to prevent the higher price (and the implied painful adjustment) from being made effective. But the attempt to shield people from the required adjustments simply shifts the burden of adjustment to others for whom the adjustment is even more costly. One does not effectively solve a problem by trying to run away from it or by shifting it to someone else.

The unwillingness to face reality—the sheer escapism—inherent in price controls has fostered all sorts of illusions about what they can accomplish. Price controls on crude oil, for example, were supposed to accomplish a number of things that they could not and were not supposed to do many of the things they did do. On the one hand, controls were intended to do the following.

1. They were supposed to reduce the prices of refined products. They did not (though the EP may have done so).
2. They were supposed to prevent windfall profits from accruing to the oil industry. They did not; they transferred the profits from producers to refiners.
3. They were supposed to insulate the United States from the effects of increased prices established by OPEC. They did not. On the contrary, together with EP, controls probably made it profitable for OPEC to set a higher price than it would have set in the absence of controls.

On the other hand, given a national energy policy of three administrations supposedly aimed at reducing oil imports, controls were presumably not supposed to increase the dependence of the United States on imported oil. Yet the controls did precisely that.

In addition, by forcing relatively high cost adjustments to changing economic conditions, price controls on crude oil resulted in substantial waste of resources. Conservative estimates place the waste caused by controls on crude oil in 1979 alone at $3 billion. Less conservative, and more realistic, estimates about the loss of domestic crude oil production owing to controls and consideration of the bureaucratic costs imposed by controls doubled that estimate.[1] Even this latter estimate does not take into account the tendency for controls to have increased the world price of oil.

In this chapter I want to explain why controls had such radically different effects from those they were supposed to have. My contention is that most of the misconceptions about the effects of controls

1. See Arrow and Kalt, *Petroleum Price Regulation*, pp. 26–27.

stem from an ancient and nearly indestructible fallacy in economics, namely, the cost of production theory of value (COPT).

THE COST-OF-PRODUCTION FALLACY

That controlling the price of crude oil and thereby reducing the costs borne by refiners will reduce the prices of refined products like gasoline seems so self-evident that to many people it is hardly open to question. Certainly, the tendency for prices to equal costs and the fact that producers change prices when their costs change do appear to confirm the idea that prices are indeed determined by costs.

However, just as our commonsense observations of the physical world sometimes mislead us, our commonsense perceptions of economic relationships can also lead us astray. Thus the insistence that since prices are determined by costs, one need not question whether crude oil price controls did reduce gasoline prices, is fundamentally of the same character as the notion that one need not bother considering whether ships sailing beyond the horizon really do fall off the edge.

Of course, five hundred years after Columbus set sail, it may seem obvious that the flat-earth hypothesis is untenable. But what is the matter with COPT? Let me mention the two main problems.

The first is that explaining prices by means of costs cannot really get us very far because, ultimately, costs depend upon the prices of the inputs and resources used to produce the commodities whose prices we set out to explain. But the prices of many resources, especially human labor and the gifts of nature, can certainly not be explained by their costs of production. So there must be another principle that determines, in whole or in part, the prices of final products.

The second, and perhaps even more serious, difficulty is that COPT is simply not consistent with basic supply and demand analysis. The inconsistency becomes manifest as soon as one tries to comprehend how controls on crude oil prices might have held down the prices of gasoline and refined products. Consider Figure 8–1. The supply curve (*SS*) represents the willingness of sellers to produce and sell gasoline at alternative prices, and the demand curve (*DD*) represents the willingness of buyers to buy gasoline at alternative prices. The higher the price (measured vertically), the greater is the amount

Figure 8-1. Supply and Demand Analysis of the Price of Gasoline.

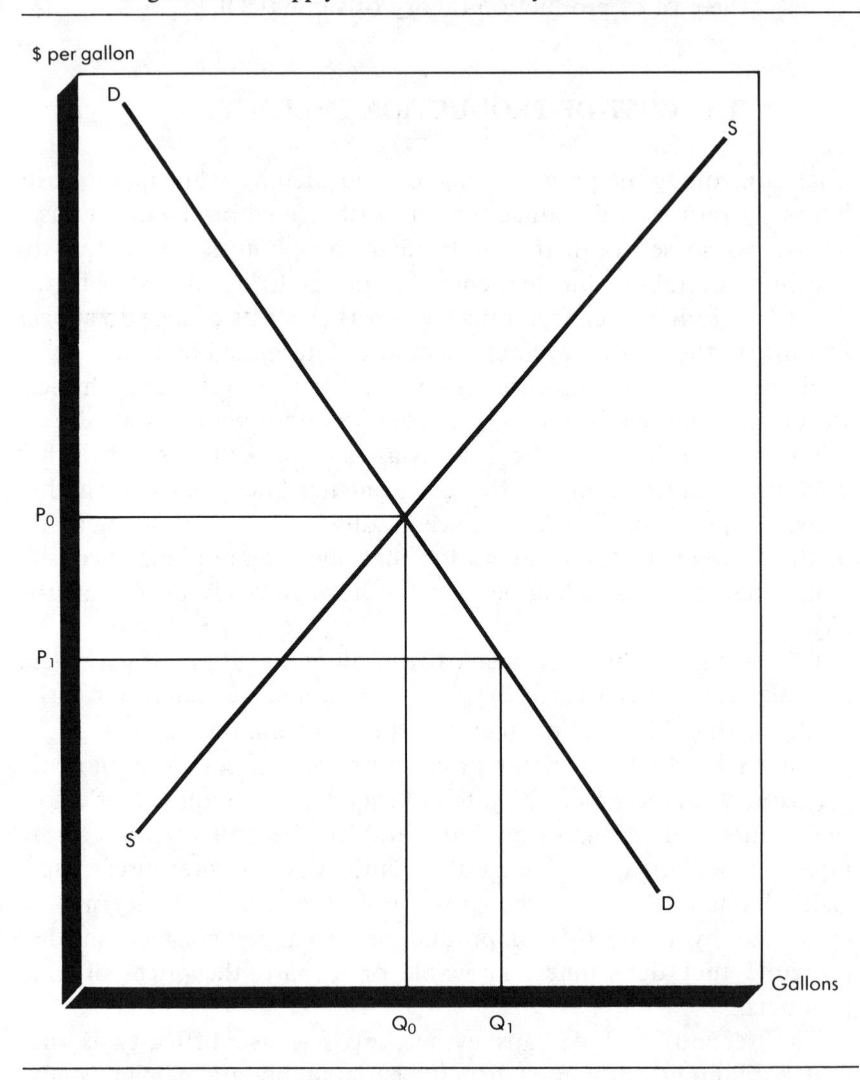

(measured horizontally) that sellers want to sell, and the smaller is the amount consumers want to buy. The intersection of the two curves determines a market-clearing equilibrium price for gasoline (P_0) at which the amount sellers want to sell (Q_0) is just matched by the amount that consumers want to buy (Q_0).

Suppose controls on crude oil prices did in fact reduce the price of gasoline to consumers to price P_1. At the lower price, consumers would have wanted to increase their consumption of gasoline to Q_1. But, unless the amount of gasoline supplied also increased to Q_1, there would have been a shortage of gasoline at price P_1.

Although shortages occurred during a few relatively brief periods under controls, on the whole consumers did manage to purchase as much gasoline as they wanted to. Thus if controlling the price of gasoline really did reduce the price as COPT suggests, the question arises: Where did the extra gasoline come from? There are only three possibilities. First, domestic oil production could have increased to allow domestic refiners to produce additional gasoline; second, additional gasoline could have been imported; and third, domestic refiners could have imported additional crude oil to produce the extra gasoline.

The first of these possibilities can be dismissed immediately. It implies that holding down the price of crude oil to domestic producers induced them to increase their rate of production. Since, without imports, domestic production could not satisfy domestic demand, increased domestic production would not have made more crude oil available to domestic refiners. It would merely have displaced an equivalent amount of imports.

The second possibility is no more plausible. Holding down the price of gasoline in the domestic market could certainly not have increased the incentive for foreign refiners to export gasoline to the United States.

As for importing more crude oil, why would a refiner increase imports of expensive crude oil just to be able to refine it and sell the products at prices that, according to COPT, did not fully reflect the price of imported oil?[2]

THE MARGINALIST THEORY OF PRICING

These considerations may make it more difficult to accept uncritically the seemingly plausible, if not inescapable, conclusion that price controls on crude oil held down the prices of refined products. But even if COPT is as problematic as I maintain, you would still be right to ask for an alternative theory. One cannot be expected to

2. One reason is that he would receive a subsidy for doing so. EP in fact did precisely this. I ignore the point for now but will return to it a bit later.

abandon a theory, however defective, without a better one to replace it. Luckily, an alternative theory does exist, and there are strong reasons for preferring it to COPT.

The essential insight of this theory, which I shall call the marginalist theory (MT) for reasons that will immediately become obvious—is that the price of a good is contingent on the most valuable use or uses (as measured by effective willingness to pay) to which anyone could apply a marginal unit of the good. The entire stock of water, taken as a unit, is enormously, probably incalculably, valuable. But 1 gallon, taken at the margin (which is where decisions and valuations in real life are always made[3]), is relatively unimportant, and its value low. This is because no very valuable use would be sacrificed if 1 less gallon of water were available. Taken as a unit, the entire stock of diamonds would certainly have a much lower value than the total stock of water taken as a unit. However, since one more diamond would satisfy (and one less would mean foregoing) a relatively valuable want, however, a diamond has a much greater value than a gallon of water.

If MT is correct, then the relationship between cost and price is just the reverse of what COPT asserts it to be.[4] Conceived of roughly as the value of the inputs used to make some products, cost is not—as we are inclined to suppose—what determines the price of whatever those inputs are used to produce. On the contrary, the value of any input is derived exclusively from the value of the products it helps to produce. You cannot give a product value just by making it with valuable resources. The value of the product is derived solely from its capacity to satisfy consumer wants. Hence resources are valuable insofar as they can be used to make products that consumers are willing to pay for. Since producing a widget uses up inputs that could otherwise be used to produce other products, the cost of producing one widget represents the value of products that are not produced because the widget is. If all costs do indeed represent the value of some foregone output, it makes no sense to assert that cost determines price.

3. The classic exposition of the idea that most, if not all, decisions—including nonpecuniary ones—involve a weighing of benefits and costs, or advantages and disadvantages, at some margin is to be found in Wicksteed, *Common Sense Political Economy*, Chapter 2. A powerful, if somewhat eccentric, exposition of the importance of the margin in decision-making has been given by Wanniski, *Way the World Works*, Chapter 3.

4. See Alchian and Allen, *University Economics*, pp. 84-87.

It is true that if the inputs required to produce widgets are mainly used to produce other products, the value of widgets, if they are to be produced at all, must not be less than the cost of the inputs used to produce them. In this limited sense, the value of widgets could be said to be determined by their cost of production. But the cost of using these inputs to produce widgets is entirely due to the fact that these inputs can be used to produce other valuable products. If they could produce no other valuable products, their value would necessarily be imputed directly from the value of the widgets they produced. That the value of output more or less corresponds to the cost of its inputs is not because the value of something is determined by its cost of production but because a difference between the value of an output and the cost of producing it attracts profit seekers who erode the difference in the process of capitalizing on it.

THE EFFECTS OF CONTROLS ON REFINED PRODUCTS PRICES ACCORDING TO MT

Let me now return to the case of crude oil and refined products. In order to produce refined products while domestic crude oil prices were under control, domestic refiners had to import between 4 and 6.6 million barrels of crude oil per day to supplement the 8 to 9 million barrels per day produced domestically.[5] The imported oil was bought at the world price, which went as high as $40 per barrel), while a substantial amount of domestically produced oil was sold subject to price controls and sold for as little as $5 per barrel.

The point here is that since refiners had to import crude oil at the world price in order to produce enough gasoline to satisfy the domestic demand for gasoline, the price of gasoline must have been high enough to make those imports worthwhile. That is, the price of gasoline must have been high enough to cover the cost of the most expensive oil that refiners were buying, otherwise refiners would have done better simply to produce less gasoline and not pay the high price of imported oil. That refiners were able to obtain some crude oil for much less than the price they paid for imported oil could not have induced them to reduce their prices. To do so would only have made it unprofitable to import sufficient crude oil to meet the public's

5. *Monthly Energy Review*, August 1982, p. 32.

demand for gasoline. But since the public's demand for gasoline was, for the most part, being met at prevailing prices (i.e., there were usually no shortages) gasoline prices must have been high enough to make it profitable to import crude oil.

I cannot leave matters resting here, however, because (as I showed in the previous chapter) competition among refiners for the subsidy implicit in the controls on crude oil prices led to the implementation of EP. EP thus redistributed the benefits of controls on crude oil prices among refiners in a politically more acceptable fashion than the distribution that would have resulted from the controls alone. But in the process of redistributing the benefits of controls, EP also affected the price of refined products. I shall now discuss those effects.

THE ENTITLEMENTS PROGRAM AND REFINED PRODUCTS PRICES

Let me first review the main features of the EP. Every month each domestic refiner was awarded entitlements (without which a refiner was not allowed to refine price-controlled oil) equal to the amount of oil it had refined that month times the percentage of all the oil refined domestically that month that was old oil (the national old oil supply ratio).[6] Thus refiners that refined a larger than average percentage of old oil had to buy the entitlements (at a price equal to the difference between the average prices of old oil and of uncontrolled oil) of refiners that refined a smaller than average fraction of old oil. These transactions made the subsidy derived by a refiner from price controls on old oil independent of the amount of old oil it actually refined.[7]

What EP did was to give a refiner additional entitlements (which had a well-defined cash value) as it refined more oil. For each additional barrel refined, a refiner received a fraction of an entitlement equal to the national old oil supply ratio. Since refiners could most easily obtain additional oil by importing it, the entitlements subsidy was also a subsidy for the importation of oil. For example, given an

6. Beginning in 1976 this ratio was adjusted to take into account the two classes of domestic oil under price controls.

7. The small-refiner bias and other features of EP prevented the distribution of the subsidy from being proportional to refiners' crude oil input.

entitlement value of say $6 (equal to the difference between the average price of uncontrolled and the average price of controlled oil) and a national old oil supply ratio of .5 (meaning that half the oil refined by refiners nationally was old oil), a refiner received a $3-per-barrel rebate (in the form of additional entitlements) for every barrel of imported oil purchased. During the life of the program, the value of the subsidy ranged between $1.50 and $6 per barrel.[8]

Since refining additional crude oil increased the supply of refined products, EP did have a tendency to reduce the prices of refined products. And since they were a necessary condition for EP to exist, price controls on crude oil were, in a sense, indirectly responsible for whatever reduction in gasoline prices that might have resulted from EP. But no advocate of price controls on crude oil that I know of was aware—or at least admitted to being aware—that insofar as controls did reduce gasoline prices they did so by making it possible to subsidize imported oil.

Even though EP had some tendency to reduce the prices of refined products, there were at least two ways in which it tended to raise product prices. Thus it is not at all certain that, on balance, EP really did reduce product prices.

One way in which EP led to higher prices was by favoring small refiners that generally produce less of the more valuable light products (gasoline and middle distillates) and more of the less valuable heavy products (residual fuel oil) from a given amount of crude oil than their larger counterparts. This small-refiner bias, which either awarded small refiners extra entitlements or relieved them from the obligation to obtain entitlements in order to refine old oil, increased the flow of crude oil to small refiners and reduced the flow to the larger, more efficient ones. The result was that EP reduced the output and raised the prices of gasoline and heating oil.

One can get some idea of how great this inefficiency was by comparing refinery yields relative to crude input to refiners in 1980, the last full year in which controls and EP were still in effect, with 1981, the year in which controls and EP were terminated. I present several such comparisons in Table 8-1. For 1981 as a whole, we see that even though average daily crude input was 7.5 percent less than in 1980, average daily refinery yields of gasoline and distillate decreased by only 1.6 percent and 1.9 percent. Much of the improvement can

8. *Monthly Energy Review*, March 1981, p. 82.

Table 8-1. Comparison of Inputs and Outputs of Refiners Before and After the End of Entitlements Program.

	Crude Oil Input to Refiners (thousands of barrels per day)	*Percent Change from Previous Year*	*Gasoline Production (thousands of barrels per day)*	*Percent Change from Previous Year*	*Distillate Fuel Oil Production (thousands of barrels per day)*	*Percent Change from Previous Year*
1978	14,739	0.9	7169	1.9	3167	-3.4
1979	14,648	-0.6	6852	-4.4	3153	-0.4
1980	13,481	-8.0	6506	-5.0	2662	-15.6
1981	12,470	-7.5	6405	-1.6	2613	-1.9
		Percent Change from First Quarter 1981		*Percent Change from First Quarter 1981*		*Percent Change from First Quarter 1981*
First quarter 1981	12,844		6412		2761	
Second quarter 1981	12,272	-4.5	6145	-4.2	2458	-11.0
Third quarter 1981	12,558	-2.2	6527	1.8	2554	-7.5
Fourth quarter 1981	12,215	-4.9	6525	1.9	2686	-2.7
First quarter 1982	11,389	-11.4	6034	-5.9	2452	-11.2
	Residual Fuel Oil Production	*Percent Change from Previous Year*				
1978	1667	-5.0				
1979	1687	1.2				
1980	1580	-6.3				
1981	1321	-16.4				

		Percent Change from First Quarter 1981
First quarter 1981	1534	
Second quarter 1981	1258	-18.0
Third quarter 1981	1232	-19.7
Fourth quarter 1981	1265	-17.5
First quarter 1982	1197	-25.2

Source: *Monthly Energy Review*, various issues.

be attributed to the elimination of many inefficient small refiners that had only survived because of the entitlements subsidy they had been receiving.[9] The effect of eliminating the small-refiner bias is also shown by the reduction in the output of residual fuel oil, which is produced in relatively large proportions by the smallest refineries. Output of residual fuel oil in 1981 was 16.4 percent less than it had been in 1980. That these changes were not part of a long-term trend towards greater refinery yields of gasoline and distillates is evident when one looks at the corresponding changes in yields and inputs in 1980 and 1979.[10]

Perhaps the most important reason why price controls and EP very likely caused consumers to pay higher, not lower, prices for gasoline and other refined products is that they increased OPEC's optimal price for oil.[11] To explain why will require a short digression on the theory of monopoly or cartel pricing.

HOW MONOPOLIES SET PRICES

Any monopoly or cartel has the problem of selecting a profit-maximizing price from a range of possible prices that it could set. The

9. Elimination of EP also caused the retirement of many marginal refineries that had only been kept profitable by the entitlements subsidy. See *Oil and Gas Journal*, 23 March 1981, p. 82.

10. One problem with the comparison is that the procedure DOE used for calculating gasoline output from refineries was changed in January 1981 in such a way as to increase reported gasoline output compared to the previous procedure. See *Monthly Energy Review*, May 1981, pp. iv–v. This would no doubt tend to improve gasoline yields in 1981 relative to 1980. I therefore also include comparisons between the first quarter of 1981—when the improvements in efficiency caused by the elimination of the small-refiner bias were, at best, only beginning to be felt—with subsequent quarters through the first quarter of 1982. These comparisons indicate that while crude input between the first and fourth quarters fell by 4.5 percent, gasoline output actually increased by 2.6 percent. Similarly, between the first quarter of 1981 and the first quarter of 1982, crude input declined by 11.3 percent, while gasoline output declined by only 6 percent. The conclusion that EP and the small-refiner bias were responsible for substantial waste of crude oil and underproduction of gasoline and other valuable refined products, and thus higher prices for those products, is unavoidable.

11. Even if EP did have a net tendency to reduce prices, it would have been counteracted to some extent by a reduction in the flow of imported refined products into the United States. In principle the reduction in imports could have completely offset whatever tendency EP had towards lower prices for refined products, but it is doubtful whether this would have been true except in markets near ports where refined products were delivered. In most domestic markets transportation costs probably prevented imports from having a substantial impact on product prices.

monopoly or cartel will increase its revenue by raising its price if the percentage increase in price is greater than the consequent percentage reduction in sales. If the monopoly or cartel can reduce its costs by producing a lower rate of output, then even if sales fell off by a percentage greater than the percentage increase in price, it could still be worthwhile to raise the price. On the other hand, if the percentage increase in sales following a price reduction were so great that it more than compensated for the lower price and the added cost of producing the extra output, it would be profitable to reduce the price.

Suppose, for example, that a firm is selling one hundred units of its output each week for $10 per unit. If it increased price by 10 percent to $11 per unit and its sales fell by only 5 percent to ninety-five units per week, its revenue would increase from $1,000 per week to $1,045 per week. But if the firm's sales fell by 15 percent ot eighty-five units per week, its revenue would decrease from $1,000 to $955 per week. It would only be profitable for the firm to increase its price if the firm saved more than $45 per week in costs by producing fifteen less units per week than before.

The optimal price thus depends critically on how responsive sales are to a given change in price. If sales are very responsive to a change in price, then unless the added costs of producing extra output are too great, a reduction in price that allows sales to expand substantially would be profitable for the monopoly or the cartel. But if sales are not very responsive to a change in price, then an increase in price is likely to be profitable since the reduction in sales would probably not offset the combination of greater revenues per unit and the savings in cost owing to the reduction in output. Thus we can formulate a general rule that whatever increases the responsiveness of sales to a change in price (what economists call the elasticity of demand) reduces the optimal price of a monopoly or a cartel, and whatever reduces the elasticity of demand increases the optimal price of the monopoly or the cartel.

It is now easy to show how EP increased the optimal price for OPEC to set. EP not only subsidized the importation of crude oil by domestic refiners – which is incredible enough by itself – but it also increased the subsidy as OPEC increased the price of oil. (The value of the subsidy, remember, was based on the difference between the average price of decontrolled oil and the average price of controlled oil.) Any increase by OPEC in the price of oil was cushioned for domestic refiners by an increase in their entitlements subsidy. Conversely, any reduction in price by OPEC was partially offset for domestic refiners by a reduction in their entitlements subsidy. This policy—for which "perverse" is surely not too strong a word—reduced the elasticity of demand for OPEC oil and thus increased the optimal price for OPEC to set. Nor can the effect have been trivial since the value of the subsidy ranged between $1.50 and $6 per barrel, and U.S. imports accounted for between 13.3 percent and 19.8 percent of total OPEC production while EP was in effect.[12]

To my knowledge, only one attempt has yet been made to estimate the effect of the entitlements subsidy on the price of oil. This study found that OPEC, as of early 1979, was charging $2.23 more per barrel with EP than it would have charged in EP's absence.[13] Studies of the effect of the entitlements subsidy on the price of gasoline that have ignored its effect on the price of crude oil have estimated that the subsidy reduced the price of gasoline by about 4¢ per gallon.[14] If the estimates are correct, the price of gasoline was most likely increased by EP, because the program actually increased the cost of crude oil. Taking the effects of the small-refiner bias into account only reinforces this conclusion.

TESTING COPT AND MT

When scientists have competing theories about how the world works, they seek to choose between the theories by deducing from them some conflicting implications or predictions about observable events. If the observations contradict the predictions of one theory and correspond to those of the other, the contradicted theory is usually rejected, and the other survives to face other theories and further tests. It is through this process of conjecture and refutation that our knowledge about the world grows and evolves. For adherents of theories to contribute to, rather than impede, this process, they must endeavor to make clear what potentially observable events would refute their theories or would be more consistent with a competing theory than with their own.

12. *Monthly Energy Review*, August 1982, pp. 36, 95.
13. Smith, "U.S. Oil Policy," p. 92.
14. Kalt, *Oil Price Regulations*, p. 174.

In this chapter I have described two competing theories about how prices in general, and the prices of gasoline and other refined products in particular, are determined. What I have been most concerned with was whether controlling the price of domestic crude oil held down the prices of refined products. COPT asserts that the controls did so, and MT (with some qualifications owing to the effects of EP) denies this. Between June 1979 and January 1981, domestic crude oil prices were gradually, but not fully, decontrolled; and in January 1981 President Reagan ordered total decontrol. The two theories would appear to have rather different implications about how gasoline prices should have behaved during and since decontrol. We thus have an unusual opportunity to test these competing theories of price determination.

Before I present evidence about what happened to gasoline prices during and after decontrol, I want to recount some of the predictions that were made about what would happen to gasoline prices in the event of decontrol. Opponents of decontrol warned, and many advocates of decontrol conceded, that huge increases in the price of gasoline and other refined products would follow the decontrol of domestic oil prices. There were warnings when President Carter started gradual decontrol in June 1979, and these were repeated before President Reagan ordered total decontrol in January 1981. And the warnings were dire indeed.

For example, in March 1980, Mr. Robert Baldwin, president of Gulf Oil Corporation's refining and marketing division (and presumably as an executive of a major oil company, a supporter of decontrol) was reported by the *Wall Street Journal* to have said: "The price of gasoline will continue to go up. I would think by Christmas it is likely to be up by another 20 to 25 cents."[15] In March 1980 the average price of gasoline as calculated by DOE was $1.230 per gallon. In December 1980 DOE reported the average price of gasoline as $1.231 per gallon. (See Table 8-2.)

In early January 1981, anticipating that President Reagan would order complete decontrol soon after taking office, Representative Edward Markey warned in the *New York Times*: "Decontrol means handing a blank check to the oil companies and asking OPEC to fill in the amount." Mr. Markey worriedly reported the prediction of the famous moderate, Sheik Yamani of Saudi Arabia, that the price of

15. *Wall Street Journal*, 21 March 1980, p. 4.

Table 8-2. Predicted and Actual Gasoline Prices.

	Predicted Prices of Gasoline				*Average Refiner Cost of a Barrel of Crude Oil*		
	Derived from Price of Imported Oil Unadjusted for EP	*Derived from Price of Imported Oil Adjusted for EP*	*Derived from Price of Composite*	*Actual Price of Gasoline*	*Imported Oil Without Adjustment for EP*	*Imported Oil With Adjustment for EP*	*Composite*
June 1979	–	–	–	$ .880	$21.03	$18.02	$17.00
July	–	–	–	.930	23.09	19.55	18.85
August	–	–	–	.967	23.98	20.20	19.75
September	–	–	–	.998	25.06	21.14	20.14
October	–	–	–	1.006	25.05	21.05	20.68
November	$1.047	$1.044	$1.030	1.019	27.02	22.63	22.04
December	1.096	1.086	1.073	1.042	28.91	24.20	23.63
January 1980	1.146	1.123	1.108	1.110	30.75	25.47	24.81
February	1.191	1.173	1.146	1.186	32.40	27.26	26.11
March	1.222	1.207	1.172	1.230	33.42	28.37	26.88
April	1.228	1.215	1.183	1.242	33.54	28.44	27.09
May	1.252	1.212	1.206	1.244	34.33	28.11	27.85
June	1.260	1.240	1.235	1.246	34.48	29.04	28.80
July	1.260	1.252	1.234	1.247	34.51	29.47	28.73
August	1.262	1.259	1.235	1.243	34.44	29.73	28.70
September	1.267	1.295	1.248	1.231	34.46	30.94	28.96
October	1.276	1.313	1.267	1.223	34.46	31.33	28.96
November	1.290	1.342	1.277	1.222	35.09	32.49	29.79
December	1.307	1.385	1.320	1.231	35.63	34.11	31.39
January 1981	1.387	1.503	1.408	1.269	38.85	38.85	34.86
February	1.396	1.513	1.472	1.353	39.00	39.00	37.28
March	1.382	1.500	1.480	1.387	38.31	38.31	37.48

April	1.388	1.507	1.463	1.381	38.41	38.31	37.48
May	1.378	1.498	1.457	1.370	37.80	37.84	36.11
June	1.363	1.484	1.436	1.362	37.03	37.03	35.03
July	1.358	1.480	1.435	1.353	36.58	36.58	34.70
August	1.344	1.466	1.433	1.348	35.82	35.82	34.41
September	1.340	1.464	1.432	1.358	35.44	35.44	34.11
October	1.341	1.465	1.432	1.358	35.43	35.43	34.07
November	1.360	1.484	1.439	1.353	36.21	36.21	34.33
December	1.356	1.450	1.441	1.348	35.95	35.95	34.33
January 1982	1.347	1.473	1.435	1.341	35.54	35.54	33.95
February	1.348	1.474	1.424	1.318	35.48	35.48	33.40
March	1.313	1.439	1.385	1.268	34.07	34.07	31.81
April	1.286	1.412	1.363	1.210	32.82	32.82	30.83
May	1.288	1.416	1.377	1.224	32.71	32.71	31.08

Sources: *Monthly Energy Review*, *Monthly Labor Review*, various issues.

oil would reach $50 per barrel and the projection of the Congressional Research Service that by spring the price of gasoline would reach $1.65 per gallon.[16]

In the January 18 edition of the *New York Times*, while advocating immediate decontrol of oil prices, Roger Stobaugh and Daniel Yergin of the Harvard Business School wrote: "Under phased decontrol, as currently working, the price of a 'composite' barrel would rise gradually and reach completely decontrolled levels by October 1, 1981. At that time gasoline at the pump would be about 22¢ a gallon higher than it is today, so long as OPEC does not increase prices further. About 10¢ would be due to the present OPEC price rise and 12¢ to decontrol. But with immediate decontrol the 12¢ rise would occur in a matter of weeks rather than months."[17]

Immediately after decontrol, the *New York Times* reported, "Consumer groups put the likely increase [due to decontrol] at 10–12¢ by the end of the summer." It also reported the charge of the New Jersey Energy Commission that decontrol "would fan the fires of inflation."[18]

As late as September 1981 adherents of COPT were forecasting increases in the price of gasoline. The *Wall Street Journal* printed an article to that effect under the headline: "Citing Surging Costs, Refiners Say Gasoline Prices Will Have to Rise."[19] In September 1981 the average price of gasoline stood at $1.358 per gallon. By January 1982 the average price had slipped to $1.341 per gallon, and by April 1982 it stood at $1.210 per gallon. (See Table 8–2.)

Despite the failure of any of their predictions to come true, adherents of COPT no doubt continue to believe that subsequent events have proven their prescience. After all, has the price of gasoline not increased substantially since June 1979 when gradual decontrol was begun by President Carter? Yes, but what of it? Given the substantial increase in the price of imported crude oil since June 1979, a simple—not to say simple-minded—comparison of the price in 1979 with the price today is quite meaningless.

At the margin, imports are the source of supplies of crude oil. It is, therefore, the price of imported crude oil that determines how much crude oil will be refined domestically and, hence, how much

16. *New York Times*, 7 January 1981, p. A–23.
17. *New York Times*, 18 January 1981.
18. *New York Times*, 29 January 1981, A–1.
19. *Wall Street Journal*, 8 September 1981, p. 37.

gasoline will be produced and sold. Thus, according to MT, although decontrol of domestic crude oil prices would have no effect on gasoline prices (apart from repercussions through EP), an increase in the price of imported oil would tend to increase the price of gasoline.

A brute comparison of the prices of gasoline before and after decontrol that ignores the effect of the increased price of imported oil is no test of COPT and MT and their predictions about the effects of decontrol. A proper test of the theories requires an explicit formulation of the relationship between crude oil prices and gasoline prices implied by the two theories.

Of course, there are those who take it as settled that decontrol must have raised gasoline prices. Even if it should be shown that prices did not rise as much after decontrol as they had predicted, to them that would simply be evidence that something else happened, say, a worldwide glut of oil that mysteriously appeared to force prices down just after decontrol. Thus any price increase that occurred after decontrol was *necessarily* the result of decontrol, while any subsequent decrease *must* have been the effect of some other cause. With rules like that, you cannot lose.

Nevertheless, COPT and MT do differ fundamentally. It should therefore be possible to deduce from them potentially conflicting predictions about the effect of decontrol on gasoline prices even if the price of imported oil was rising.

The following example will show how conflicting predictions about the effects of decontrol on gasoline prices can, in fact, be derived from the two theories. Suppose that one fourth of all the oil refined is subject to price controls and must be sold for $10 per barrel, while uncontrolled oil sells for $30 per barrel. If price controls were removed, COPT would predict that the price of gasoline would increase by about 12¢ per gallon (from $1.00 to 1.12 per gallon, for example). This is because the average price of a barrel of oil would increase by $5 from $25 to $30 per barrel. Since there are 42 gallons per barrel, $5 per barrel corresponds to 11.9¢ per gallon.

By contrast, MT holds that even before decontrol the price of gasoline must have reflected the $30-per-barrel price of uncontrolled oil. Otherwise it would not have paid refiners to import oil at $30 per barrel to produce additional gasoline. But we must also take EP into account if we are to make any prediction about the effect of decontrol on gasoline prices. Under my assumptions, the price of an entitlement would be $20. If the national old oil supply ratio were

25 percent, then for every barrel of oil a refiner imported, it would receive .25 entitlements, or the equivalent of $5. The effective price a refiner paid for imported oil would only be $25 per barrel—not the $30-per-barrel nominal price. If so, MT and COPT would (though for different reasons) seem to yield the same predictions about the effect of decontrol on the price of gasoline.

It is not, however, quite so simple. First, note that if the old oil supply ratio were less (or more) than 25 percent, the subsidy to imported oil would be less (or more) than $5 per barrel; and MT accordingly would imply a smaller (or greater) price increase following decontrol than would COPT. Second, MT recognizes that an increase in the price of crude oil (at the margin) may be reflected in diminished prices for other inputs and not entirely in an increased price of gasoline. In Chapter 6 I explained why a reduction in the price of refining equipment was, in fact, a likely consequence of an increase in the price of crude oil. The removal of the $5-per-barrel entitlement subsidy only implies an upper bound to the increase that could be expected according to MT. A smaller increase would not contradict MT.[20] But according to COPT the predicted increase cannot be less than the full amount of the increase in the average price of crude oil.

Suppose that while oil prices were decontrolled, the price of uncontrolled oil increased from $30 to $35 per barrel. Based on the $10-per-barrel increase in the average price paid for crude oil, COPT would predict an increase of about 24¢ in the price of gasoline to about $1.24 per gallon.

According to MT, if the national old oil supply ratio were 25 percent, the increase in oil prices at the margin would be $10 per barrel (the loss of the $5 per barrel subsidy associated with EP and the $5-per-barrel increase in the price of imported oil). Thus MT and COPT would seem to yield identical predictions about the effect of decontrol on gasoline prices, but the same qualifications I mentioned just a moment ago apply here as well. While COPT would predict an increase of 24¢ per gallon, MT only predicts that the increase would not exceed 24¢ per gallon. In our example inputs other than crude oil were receiving at the margin 40.5¢ per gallon out of the $1.00 per gallon price of gasoline before decontrol. A 24¢-per-gallon increase in the price of crude oil could simply be passed along to con-

20. See next footnote.

sumers in the form of higher prices. That is what COPT asserts *must* happen. But market conditions could compel other inputs to settle for less than the 40.5¢ per gallon they had been receiving so that the price of gasoline would rise by less than the full 24¢ per gallon.[21] MT recognizes this possibility, which is not even considered by COPT.

Thus if COPT were true, the increase in gasoline prices following decontrol would be equal to the per-gallon increase in the average price refiners paid for crude oil. But, according to MT, when some of the domestic crude oil refiners purchase is under price controls, the price of gasoline could rise by no more and might well rise by less than the per-gallon increase in the price (adjusted for the entitlements subsidy) refiners have to pay for imported oil.

RESULTS OF THE TEST

What I have done to test the theories empirically is simply to use the average price refiners pay for crude oil from all sources and the average price they pay for imported oil (adjusted for the entitlements subsidy) to predict the price of gasoline. Such a test could have several possible outcomes. (1) The predicted prices derived from the average price of crude oil could exceed those derived from the price of imported oil. (a) If actual prices equaled or exceeded those derived from the average price of crude oil, MT would be falsified. (b) If actual prices were below those derived from the price of imported oil, COPT would be falsified. (2) The predicted prices derived from the average price of crude oil could be below the upper bound for gasoline prices derived from the price of imported oil. In that case, it would be more difficult to falsify one of the two theories. (a) If prices were substantially below those predicted by COPT, it would cast serious doubt on the validity of COPT but not on the validity of MT. (b) If actual prices were significantly above the upper bound predicted by MT, that would cast greater doubt on MT than on COPT.

21. To what extent an increase in the price of crude oil can be passed forward to consumers instead of reducing the price of cooperating inputs depends on the elasticity of demand for products, that is, the elasticity of substitution; and the less elastic the supply of other inputs, the more likely it is that an increase in the price of crude oil will reduce the prices of cooperating inputs rather than raise the prices of refined products. See my discussion in Chapter 6, "How Ceiling Prices Could Be Above Marketing-Clearing Prices."

Predicted prices for both COPT and MT can be calculated easily because every month DOE and its predecessors computed refiners' average acquisition costs of imported crude oil and their average acquisition costs of a composite barrel of oil, that is, the weighted average of all prices paid by refiners for crude oil from all sources. Since DOE also calculated the magnitude of the entitlements subsidy for importing oil while EP was in effect, we can adjust the cost of imported oil to domestic refiners by the amount of the entitlements subsidy. In addition, DOE computed the average pump price of gasoline for all grades each month.

Using DOE estimates, I derived predictions for average gasoline prices for each month starting with November 1979. Although decontrol began in June 1979, I started with November because gasoline and crude oil prices (apart from decontol) had been stable in September and October 1979. This was not true of any two consecutive months for some time before or after. The stability suggests that the relationship between crude oil and gasoline prices during those months was not too far from an equilibrium and that October 1979 is a good starting point for deriving predictions about prices in subsequent months.

To derive predictions about the price of gasoline, one must also take into account the costs attributable to noncrude inputs used in refining, transporting, and marketing gasoline. For MT, such costs have limited predictive value since to some extent they do not determine the price of gasoline but rather are determined by it. According to COPT, however, the costs of noncrude inputs should affect the price of gasoline quite as much as the cost of crude oil does.

The costs of refining, transporting, and marketing gasoline that are attributable to inputs other than crude oil can be inferred from the difference between the price of gasoline and the per-gallon cost of crude oil. But how should the cost of crude oil be measured? According to COPT, we should measure it by the cost of all the crude oil that is used or, in other words, by the cost of the composite barrel of crude oil. Thus COPT would measure the cost of noncrude inputs by taking the difference between the price of gasoline and the per-gallon cost of the composite barrel of crude oil. According to MT, we should measure the cost of crude oil at the margin or, in other words, the cost of imported crude oil. Thus MT would measure the cost of noncrude inputs by taking the difference between the price of gasoline and the per-gallon cost of imported oil adjusted for the entitlements subsidy. The former estimate for October 1979 works out to

51.4¢ per gallon; the latter estimate for October 1979 works out to be 50.5¢ per gallon. Without the adjustment for the entitlements subsidy, the estimate would be 41¢ per gallon.

I calculated the predicted price of gasoline in any month by taking the relevant per-gallon cost of crude oil in that month and adding to it the approximate noncrude oil costs (50¢ per gallon) multiplied by an inflation adjustment factor. This factor was the ratio of the consumer price index in any succeeding month to its value in October 1979. The monthly predictions of the two theories about the price of gasoline along with the actual average price of gasoline as calculated by the DOE are presented in Table 8–2 and Figure 8–2. Table 8–2 and Figure 8–2 also present predictions calculated from the price of imported oil with no adjustment for the entitlements subsidy.

Two main results emerge from Table 8–2. (1) COPT predictions derived from the cost of a composite barrel exceed actual prices in each month except from January to August 1980. (2) MT predictions derived from the price of imported oil adjusted for the entitlements subsidy are generally a few cents above the COPT predictions.

Although COPT predictions seem to be more accurate than the MT predictions, one must remember that COPT predictions were almost uniformly above actual prices and that while, according to COPT, prices should not be less than predicted prices according to MT, predicted prices are merely upper bounds and would be quite likely to exceed actual prices.

Interestingly the predicted prices derived from the price of imported oil unadjusted for EP are easily the most accurate of the three sets of predictions. This confirms that the entitlements subsidy largely accrued to inputs in the production of gasoline other than crude oil.

A crucial period is the one from January to March 1981, when crude oil prices and gasoline prices were all rising very rapidly and when all domestic crude oil prices were completely decontrolled. Opponents of decontrol and adherents of COPT have cited the increases in the price of gasoline following President Reagan's announcement of decontrol on January 28 as proof that, just as they had predicted, decontrol did lead to substantial increases in gasoline prices. The evidence presented in Table 8–2, however, supports a very different explanation for these increases.

Consider, first, the period from March to November 1980. During this period ample supplies of refined products kept prices stable. In October 1980 war broke out between Iran and Iraq. Because actual

Figure 8-2. Predicted and Actual Prices of Gasoline.

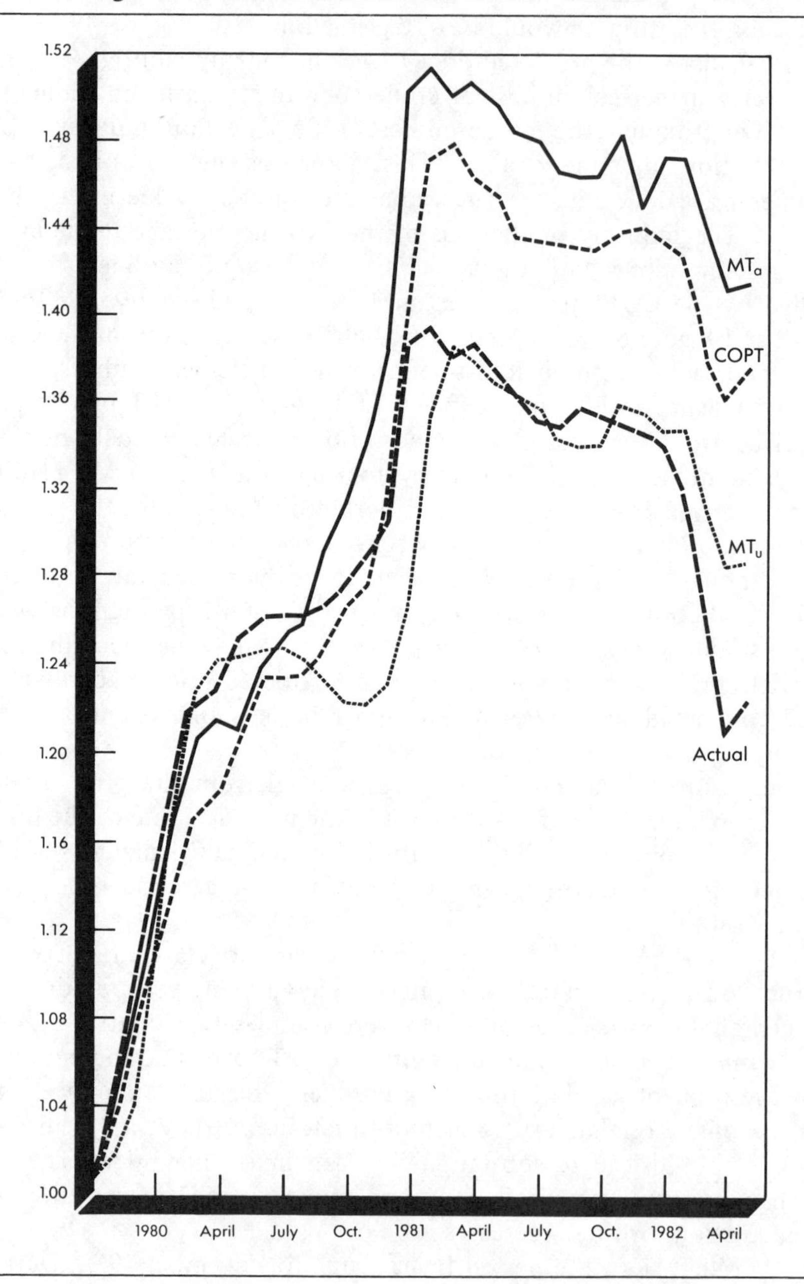

prices seem to have been low relative to the prices for imported crude oil, had these two major oil producers not gone to war with each other, crude oil prices might well have fallen. The outbreak of war changed all that. Given the unusually large inventories that had been accumulated in the previous year and a half, the war at first did little more than relieve downward pressure on, and allow for some modest increases in, crude oil and product prices. But by January, when prospects for a quick end to the war were dashed and fears of a Saudi production cutback gained currency, the price of crude oil began rising rapidly in world markets.

Based on the increase in the price of imported crude oil in January 1981, and without adjustment for the effect of EP, which was terminated that month, the predicted price of gasoline rose by 8¢ per gallon to $1.387. Taking termination of EP into consideration implies a predicted increase of almost 12¢ per gallon. But the average price in January actually rose by only 3.8¢ per gallon to $1.269. The differences between actual and predicted prices derived from the prices of imported oil are larger in January than for any other month for which I derived such predictions. Subsequent predictions in the price of gasoline can therefore very plausibly be explained as belated adjustments to increases in the price of imported oil that had already taken place rather than as the effects of decontrol. Table 8-3 shows that prices had started to increase rapidly before decontrol was announced, and it is hard to believe that the trend would not have been essentially the same had decontrol not been ordered.

The superiority of MT over COPT in explaining gasoline prices is, I think, quite clear. The price increases that actually occurred are

Table 8-3. Average Pump Price of Regular Gasoline.

December 30, 1980	$1.234
January 7, 1981	1.244
January 14, 1981	1.254
January 21, 1981	1.270
January 28, 1981	1.286
February 4, 1981	1.328
February 11, 1981	1.337
February 18, 1981	1.355
February 25, 1981	1.357
March 4, 1981	1.363
March 11, 1981	1.364
March 18, 1981	1.362

Source: *Oil and Gas Journal*, various issues.

substantially less than those implied by COPT. Although the actual increases are also substantially less than the *maximum* increases implied by MT, the fact that the increases turned out to be less than the maximum prediction is itself suggested by MT.

Moreover, as I pointed out earlier, the superiority of MT has still another dimension since MT predicted—while COPT failed to predict—the decrease in crude oil prices following decontrol. Adherents of COPT invoked the decrease in oil prices following decontrol to save their theory from falsification. *Newsweek* summed up the COPT position succinctly: "Decontrol sent prices soaring. Now a global glut has caused prices to drop."[22] But even at current oil prices, gasoline prices are not as high as they should be if COPT were correct. So COPT cannot be saved by attributing to the oil glut the failure of gasoline prices to rise as much as COPT predicted. Nor can COPT begin to explain, as MT can, how a global oil glut mysteriously emerged just a few months after "experts" like the moderate and all-knowing Mr. Yamani of Saudi Arabia had been predicting that the price of oil should soon rise to $50 per barrel and just as OPEC oil output was plunging to a ten-year low.

As I have already explained, MT did predict that the price of crude oil would decline as a consequence of decontrol because elimination of the entitlements subsidy to imported oil would increase the demand elasticity for OPEC oil. By 1981 the implied reduction in the world price of oil was probably well over the $2.23 that was calculated for 1979, perhaps as high as $4.00 per barrel.[23] As one can see from Table 8-2, the average cost of crude oil to refiners fell by about $6 per barrel in the year following decontrol.[24]

22. *Newsweek*, 13 July 1981, p. 57.

23. Using data for September 1980, I recalculated Smith's estimate of the increase in the world price of crude oil caused by EP. My calculations showed the increase to be between $3.81 and $4.34 per barrel. The range in my estimate reflects an allowance for the possible change in the elasticity of demand for OPEC oil.

24. Decontrol also operated through another less direct mechanism to reduce crude oil and products prices. The end of the entitlements subsidy permanently reduced the level of oil imports to the United States below the levels it would have attained had controls remained in effect. Under any given monetary policy, this implied an increase in the dollar's exchange rate with foreign currencies relative to its exchange rate had controls remained in effect. As foreign currencies became cheaper relative to the dollar, oil prices in terms of those currencies increased at the prevailing price of oil quoted in dollars. Since oil prices in international markets are always quoted in dollars, foreign consumers experience higher oil prices as a result of decontrol in America. The consequent reduction in foreign oil consumption intensified the oil glut and increased the downward pressure on world oil prices.

CONCLUSION

Price controls on crude oil had virtually none of the beneficial effects its proponents attributed to them. Consumers were not made better off; the oil industry as a whole was not prevented from profiting on the rising world price of oil; and the United States economy was not cushioned from the effects of rising world oil prices. Indeed, the controls along with EP probably led to higher world and domestic prices for crude oil and higher products prices for consumers than would have prevailed in their absence. Although controls did not achieve their stated public policy objectives, they clearly did serve the interests of various groups, principally—but not exclusively—small refiners, that used the controls as a means for maintaining an off-budget subsidy program.

In the next two chapters, I shall show how similar misconceptions have led to the windfall profits tax on crude oil and to controls on natural gas prices. These measures have also failed to achieve their stated objectives, but they have afforded opportunities for certain interests to exploit the programs for their own ends through the political process.

Chapter 9

WINDFALL PROFITS AND WHAT TO DO ABOUT THEM

I put it to you that we shall never see President Carter—or any other President of the United States—embrace and indeed plant a kiss on the cheek of the chief executive officer of a major oil company, even though we have already seen him embrace and kiss the chief executive officer of what is now the greatest tyranny on the face of the earth. Secretary Vance has said that Mr. Carter and Mr. Brezhnev "share the same dreams and aspirations;" he would never be so impolitic as to say that of Mr. Carter and the president of Exxon Corporation, even though in this case it is probably true.

Walter Berns*

INTRODUCTION

In the preceding chapter I attempted to do two things; first, to explain in theory why controls on crude oil prices did not hold down the prices of refined products to consumers and, second, to present evidence in support of the theory. In presenting this evidence, I showed that increases in gasoline prices during and shortly after the decontrol of crude oil prices were attributable to rising prices of imported crude oil, not to decontrol itself.

*The epigraph is an excerpt from testimony prepared for Congressional hearings on Senator Kennedy's Energy Antimonopoly Bill and was quoted in *The Wall Street Journal* 31 October 1979, p. 22.

Holding down prices to consumers was certainly one very important objective of controls on crude oil prices, but I want to recall another stated objective of the controls, simply to keep down oil company profits. The two objectives were intimately related, of course, since the ostensible function of the controls was to transfer income from the oil companies to consumers by forcing the former to charge lower prices to the latter. My point is that reducing oil company profits achieved the status of an independent goal of public policy and was not merely a means to achieving other desired goals. The oil companies had become the objects of so nearly universal antipathy that almost any measure, regardless of its other consequences, could count on a Congressional majority in its favor if it promised to reduce oil industry profits. The irresistible attraction of controls on crude oil prices was that, in promising to reduce the profits of the oil companies along with product prices, they appealed simultaneously both to the public's hostility to the oil industry and to the public's own notion of self-interest.

If, as I have shown, controls on crude oil prices did not reduce the prices consumers were paying for refined products, it follows that the profits of the oil industry as a whole (as distinguished from particular companies) were not reduced either (or at best were only reduced because of inefficiencies that did not benefit consumers). Instead of transferring income from oil producers to consumers, the controls on crude oil prices transferred income within the oil industry from producers to refiners. This intraindustry transfer was crucial to forming a viable coalition in favor of controls. Without this transfer, consumer antipathy towards the oil companies and their desire for lower prices might well have been too ephemeral and too diffuse to have sustained controls on oil prices in the face of the unified opposition of the oil industry.

In general a coalition made up of a relatively small number of similar members, each of whom has a fairly substantial interest at stake, is likely to be more successful in political competition than another coalition made up of a larger number of heterogeneous members, each of whom has less at stake than members of the former coalition. This would be true even if the latter coalition as a whole stood to gain or lose more than the former from the policy at issue. The reason is that each member of the former coalition individually gains more from a favorable outcome and has a greater impact on the final outcome than does a member of the latter. Thus members of the former coalition have more incentive to expend resources in competing politically than do members of the latter.

This explains why, for example, tariffs continue to be imposed on imported goods even though it has been demonstrated to the satisfaction of almost every economist who ever lived that the gains to producers from the tariffs are more than offset by the losses imposed on consumers. Nevertheless producers are usually better able to advance their common interests than are consumers because the former are fewer in number than the latter and because each producer has more at stake than does any consumer. Producers therefore have a far greater incentive to devote resources to achieve the imposition of tariffs than consumers have to resist them. The same reasoning applies to any number of situations in which legislation or regulation is turned to the advantage of a small well-organized group and to the disadvantage of the general public.[1]

Although the intraindustry transfer may have been the key to maintaining a dominant coalition in support of controls, the controls did more than just transfer income within the industry. As I explained in Chapter 8, controls gave rise to the EP. On the one hand, the EP tended to lower product prices by subsidizing the importation of refined products; on the other hand, it tended to increase product prices by subsidizing inefficient refiners and, more importantly, by inducing OPEC to raise the price it set for oil above the level it would have imposed in the absence of controls and EP. On balance, therefore, controls and the EP probably implied a transfer not only from domestic producers to domestic refiners, but also a transfer from domestic producers and domestic consumers to domestic refiners and foreign producers.

It seems somewhat paradoxical that domestic consumers, insofar as they were exerting any influence on the formation of policy, would have done so in behalf of foreign producers and in opposition to their own interests. Of course, it cannot be excluded a priori that the antipathy of domestic consumers toward domestic oil producers was so great that they would have chosen, in full awareness of the effects, to penalize the oil industry even though they would, as a result, have to pay higher prices. What is more difficult to understand is why consumers would have sought to penalize domestic oil pro-

1. George J. Stigler, (*The Citizen*) has elucidated this point with great vigor, clarity, and wit.

ducers in a manner that redounded to the benefit of foreign oil producers for whose benefit consumers, one would have thought, would not have been quite so eager to make sacrifices. All this suggests that the role the general public played in the formulation of oil policy was played in ignorance of the effects the policy was actually having on its interests.

George Stigler has told us that government policies cannot simply be explained away as mistakes.[2] By this he means that government policies are the results of the purposive behavior of rational individuals and groups who presumably are getting something valuable in return for their efforts. I should not differ with that as far as it goes. Indeed, to say that the outcome of such activities is wholly unintended and mistaken is not only implausible, it is unenlightening. But ignorance is certainly one fact that economists must reckon with in their analysis of market behavior, and they should not be less willing to acknowledge its role in the political process than they are in markets for goods and services. Just as the exploitation by some people of others' ignorance results in continuing wealth transfers in markets for goods and services (witness the continuing sales of cures for baldness and laetrile treatments for cancer), the exploitation of ignorance facilitates wealth transfers through the political process.

With the removal of price controls on crude oil, the wealth transfers that had been effected by the controls were replaced by a new set of wealth transfers imposed by the windfall profits tax (WPT). WPT was designed to prevent the oil industry from being enriched by decontrol, but since the main effect of decontrol on the oil industry was to terminate an intraindustry transfer, the oil industry as a whole gained little from decontrol. Indeed, Mobil Oil—a major company with relatively small domestic oil reserves that had been a net entitlements seller—actively supported the retention of controls instead of decontrol with WPT.[3] Thus instead of transferring the wealth extracted from domestic oil producers to domestic refiners, as controls had done, WPT transferred that wealth to the federal Treasury and from there to a variety of groups such as producers of synthetic fuels and energy conservation devices. Foreign oil producers continued to partake in a share of the wealth extracted from domestic oil producers as well.

2. See Stigler, *The Citizen*, and *Economist as Preacher*.
3. See *The Wall Street Journal*, 4 May 1979, p. 10.

That WPT could be enacted to replace the controls on crude oil prices even though WPT implied a lower after-tax income for the oil industry than they had earned under price controls shows how important a goal reducing oil industry profits had become. Just the appearance that the oil industry's profits would rise because of decontrol was too offensive to be tolerated.

The moral justification of WPT is an issue I shall discuss later in this chapter, but I must first discuss some mundane questions about both price controls and WPT such as their effects on oil output, who gained and who lost from them, and the causes that give rise to windfall profits in general and within the oil industry in particular.

PRICES AND PRODUCTION

In my discussion of the effects of price controls on crude oil in earlier chapters, I spoke of the effects controls had on production only briefly and in passing. One reason for not discussing the effects of controls on production at greater length was that in the debate over controls, opponents of controls had tended to emphasize the disincentives to production. The claims of greatly increased output if controls were lifted were, of course, vigorously denied by the proponents of controls. It was my purpose to show that the case against controls could be made powerfully enough even if proponents were correct in asserting that higher prices would not elicit any substantial increase in production. Thus nearly every argument I made in demonstrating the ill effects of controls was valid even under the assumption that controls were not discouraging production. If controls also discouraged production, as opponents of controls maintained, the case against controls is even more overwhelming.

Since decontrol did away with many of the inefficiencies I discussed earlier even with WPT, the impact of controls on production must now be taken into account if any reasoned judgment is to be rendered about the effects of WPT. Normally one could take it for granted that controls or taxes that reduce the effective price received by the producer also reduce output. That much is obvious from freshman economics. But in the case of oil production, the analysis is not so simple for two reasons. First, the amount of production depends not only on the current price but also upon the relationship between the current price and anticipated future prices. Second, there seems to be an absolute limit imposed on oil production by the amount of oil that is beneath the ground. If higher prices cannot add to the amount of oil that nature has left there for us, how can an increase in price increase oil production?

Let me discuss both of these points. I explained in Chapter 2 that the rate at which a given quantity of oil would be depleted over time if the oil were costless to extract would depend only on the rate at which the oil appreciated. The level of the price trend would induce the public to consume the amount of oil that was available. In that situation it is true that increasing the level of prices could not increase output. But if there are various sources of oil, and it is costlier to extract oil from some sources than others, a higher price for oil will make it profitable to extract oil from some sources that it would not have been profitable to extract otherwise.

This is not to deny that the expected rate of appreciation in oil prices can have a major effect on the rate at which oil is extracted from any given source. Thus if the current price increased, but prices in the future were expected to rise even more rapidly, there would be some reduction in the current rate of production in order to postpone extraction until prices had in fact risen as much as anticipated. Similarly a reduction in the current price could be associated with an increase in the current rate of output if anticipated price reductions in the future induced producers to increase extraction now to avoid selling at even lower prices in the future. In either case, however, the change in price over the long term would ultimately reverse the initial effect.[4]

For the second point, that oil production is limited by the amount of oil underground rather than by the price paid to producers, reflects a widespread belief. If there is only so much oil out there to be found, it is often argued, how can raising the price increase the amount of oil produced? The answer, of course, is that the amount of oil to be found is not fixed, nor is the fraction of the oil found that will ultimately be extracted fixed. The rate at which new reserves are found or developed is actually highly sensitive to price. For thousands of years of human history, almost no oil was found. If the price of oil had remained at the negligible levels oil had commanded

4. The discussion in the text is somewhat oversimplified since, over a range of extraction rates, marginal cost is more or less constant. Beyond this range, increasing the rate of extraction is associated with rapidly increasing marginal costs. It is this fact that accounts for the relatively small response of oil output to price in the short run.

for millenia, hundreds of billions of barrels of oil should never have been found, much less extracted. It is inconceivable that all the oil that could be found ever will be found; but the higher the price of oil rises, the more oil that will be found. And the higher the price, the greater the fraction of the oil found that it will be worthwhile to extract.[5]

Suppose, for example, that a firm has discovered an offshore field with potentially recoverable reserves of 100 million barrels. To bring the field into production requires investment in all sorts of production equipment and ongoing expenditures on materials and labor to operate and maintain the production equipment. Based on estimates of feasible rates of production and the required investments and operating costs, the company will calculate a price at which it can expect production to generate enough revenue to make the entire project profitable. If the company anticipates actual prices below this magic number, it will abandon or postpone development of the field. But if the price is anticipated to be above the critical value, the company will proceed to develop the field. Of course once the investment is made, it is a case of bygones are bygones, and production would continue even if the price were to fall to a fraction of what would have made development profitable in the first place.

Since offshore fields require larger fixed investments and have higher production costs than do onshore fields, the former have to be considerably larger than onshore fields to become commercially viable. The development costs of a larger field can be spread over a larger number of barrels than the more or less equal development costs of a similarly situated, but smaller, field. Since the investment required to bring Alaskan or North Sea oil into production is far greater than that required to bring oil from the Santa Barbara Channel or the Gulf of Mexico into production, the fields in Alaska or the North Sea must be far larger than those off the Santa Barbara Channel or the Gulf of Mexico in order to be commercially viable. Obviously as the price of oil increases, the minimum size required for these fields to be considered commercially viable diminishes, while the incentive to explore for oil in these and other more remote areas increases.

The effects of price controls or of WPT on crude oil production can now be brought into clearer focus. The immediate effect of con-

5. Simon, *The Ultimate Resource*, Chapter 7.

trols was perhaps ambiguous. Although there was probably a small short-run reduction in output because of controls, an expectation that controls would become more rigorous in the future could perhaps have induced a temporary increase in output. But over the long run, the controls undoubtedly reduced output by reducing both the profitability of investment in enhanced recovery techniques and the incentive to explore for new reserves. Even though new oil was exempt from controls until 1976, the expectation that controls or WPT might later be imposed undoubtedly discouraged exploration and so reduced output below attainable levels.

The strength of the relationship between oil prices on one hand and exploration, development, and production on the other is brought out by the data on drilling activity, reserves, and oil production during the last decade. Reserves and production began to decline steadily after 1970.[6] And drilling activity was lower in 1973 than in any year since 1950.[7] But the surge in oil prices in 1973-74, although restrained by controls, did reverse the decline in drilling activity. Increased drilling and exploration succeeded in stabilizing domestic production and reserves by the end of the decade. Between 1973 and 1979 the number of rotary rigs in operation in the United States almost doubled, and by 1981 the number doubled again. Similarly the decline in oil prices that began in the spring of 1981 resulted in a substantial decline in the number of rigs in operation during 1982 and 1983.[8] It would be absurd to believe that the increase in drilling activity was not responsible for reversing the decline in domestic reserves and oil production. To do so would be to suppose that the costs voluntarily incurred for drilling and exploration had no relation to the returns from that activity.

It is sometimes argued that higher prices could not have elicited any more drilling because bottlenecks—such as shortages of drilling equipment—not profitability, was the constraint on production. Allowing the oil companies to receive higher prices for oil, it is alleged, would not have led to the discovery of any more oil but would have just increased the already overblown profits of the oil companies. Yet the whole point of a price system that communicates the market values of products and resources is to alert people to the

6. *API Petroleum Data Book*, Section II, Table 2.
7. *API Petroleum Data Book*, Section III, Table 2.
8. *Monghly Energy Review*, May 1983, p. 58.

existence of bottlenecks and to provide incentives to overcome them. Socialist economies are overwhelmed with bottlenecks, not because of absolute physical limitations on the availability of resources, but because no one stands to gain (that is, make a profit) by eliminating the bottlenecks. No bottleneck can long endure if those who eliminate it can earn a profit.

Another argument of the same type as the previous one is that oil industry profits were so great even under controls that surely no further incentive was required to search for and produce additional oil. Higher prices would have only meant higher profits, not increased output. This argument simply denies that oil companies engage in any kind of rational estimation of costs and benefits in deciding whether to undertake a potential exploration or development project. If they do rationally evaluate such projects, clearly some projects would appear profitable at a sufficiently high anticipated price but not at lower prices. In that case, some projects were undoubtedly either not undertaken at all or were delayed because price controls or WPT held down the effective price of crude oil to producers.

Although estimates of the amount by which domestic oil production has been discouraged under EPAA or EPCA price controls and under WPT vary, it is highly probable that the effects were at least in the hundreds of thousands of barrels per day and perhaps as much as one million barrels per day.[9] This loss of output, which implied a corresponding transfer of wealth to foreign oil producers, was a pretty steep price to pay just to indulge our taste for punishing the oil companies.

WPT AND ITS PROVISIONS

WPT established three categories of crude oil subject to different rates of taxation: tier-one oil, tier-two oil, and tier-three oil. The basic approach is to define a base price for each category of oil and then to tax the difference between the base price and the actual market price. The base price more or less corresponds to the level of oil prices at the time the process of decontrol was formally begun by President Carter. It is thus aimed at taxing away from oil producers

9. Kalt, *Oil Price Regulations*, pp. 287-88.

a substantial portion of the appreciation of oil that they would otherwise have enjoyed under decontrol.

Tier-one oil includes production that had been classified as either upper tier or lower tier oil under EPCA. The base price for tier-one oil is approximately the upper tier ceiling price for crude oil in May 1979 adjusted for inflation. WPT applies a 70-percent tax on the difference between the selling price of tier-one oil and the tier-one base price.

Tier-two oil includes stripper oil and oil produced from naval petroleum reserves. The tier-two base price equals the adjusted May 1979 ceiling price plus $1 per barrel. The difference between the selling price of tier-two oil and the tier-two base price is taxed at a 60-percent rate.

A special tax rate was applied to the tier-one and tier-two oil produced by independent producers (defined as firms with sales below $1.25 million per quarter or less than 50,000 barrels per day refining capacity). The first thousand barrels per day of tier-one and tier-two oil produced by such firms are taxed at only a 50-percent rate.

Tier-three oil consists of production from properties that came into production after 1978. The base price for tier-three oil is the May 1979 ceiling price plus $2 per barrel. The difference between the base price and the selling price of tier-three oil is subject to a 30-percent tax. Independent firms pay no tax on the first one thousand barrels per day of tier-three production. WPT is to remain in force until the latter of January 1988 or the first month after which $227 billion will have been collected from the WPT (but in no case later than January 1991).[10]

The reduction of the tax rate for independent firms on their first thousand barrels per day of production resulted in very little besides a reduced tax burden for those firms that qualified for the lower rate. The key point is that the lower rate applied only to the *first* thousand barrels per day so that no firm producing more than a thousand barrels per day would have any reason to change its behavior so as to take advantage of the lower rate.

Does the exemption for independent firms tell us anything about the political forces shaping U.S. oil policy? I believe so. For one thing it further demonstrates that the oil industry cannot be viewed as a monolith. The industry encompasses numerous and conflict-

10. Kalt, *Oil Price Regulations*, pp. 19-22.

ing interests that exert considerable, though frequently opposing, influences on the political process. The major oil companies, on one hand, were expecting to recover some of the revenue lost to WPT in the form of subsidies for synthetic fuel production that would be financed by those revenues. The smaller producers, on the other hand, were less likely than the majors to obtain subsidies for synthetic fuel production and required a different form of relief from WPT, that is, the lower tax rate for independent oil firms.

One also observes that small refiners, which had been deprived of their entitlements subsidy as well as the small-refiner bias that went along with it, were able to obtain a share of the special benefit that accrued to small producers. Even if we can explain why independent firms were to receive special benefits, it is still puzzling why they were not at least given benefits that would have elicited added production. Perhaps it was thought that providing a lower tax rate to independent producers at every level of output would encourage larger firms simply to split up to capitalize on the tax break. This would have prevented independent producers from excluding the larger firms from a share in the tax break of limited size that they wished to preserve for themselves.

WPT thus preserved the supply-side inefficiencies and disincentives for production that had characterized controls on crude oil prices while adding some new ones of its own. Producers of lower-tier oil did receive substantially higher prices under WPT than under EPCA controls, but the amount of lower tier oil had been declining anyway as old wells were gradually being depleted. Producers of upper tier oil were allowed somewhat higher prices under WPT than under EPCA, but the difference was insubstantial. Moreover, much oil that had been exempt from price ceilings under EPCA, such as Alaskan oil, stripper oil, and newly discovered oil, was subject to a lower effective price ceiling under WPT than under EPCA. Net prices received by oil producers therefore were lower for some categories of oil under the WPT than they had been before decontrol under EPCA.

Not only did WPT reduce incentives to produce, its multiple price ceilings also distorted incentives so as to generate further inefficiency and loss of output. The reason for this is quite simple. Output from any source will generally be increased up to the point at which the marginal cost of production from that source is equal to the price of output from that source. Efficiency requires that the marginal cost of producing the same product be equalized from the various sources of that output. Thus if you have two oil wells, and it costs you $5 per barrel at the margin to produce from one well and $10 per barrel at the margin to produce from the other, you are throwing at least $5 down the drain because you could produce one less barrel from the second well saving $10 and produce one more barrel from the first well at a cost of only $5. Had you done so, you would have the same number of barrels of crude oil and $5 extra in your pocket. If the marginal costs of production from the two wells were still not equal after production from the more costly well was reduced and increased from the less costly well, your savings could be increased still further.

With multiple price ceilings, under both price controls and WPT, output from various sources generated different marginal revenues depending on which price ceiling was applicable. Thus output from the various sources would be adjusted so as to equate the marginal cost of output from each source with the applicable price ceiling (i.e., marginal revenue) from each source. Since the price ceilings were different for different sources, the marginal costs from the different sources could not be equalized, implying a waste of resources and a further reduction of output.[11]

A NOTE ON SYNTHETIC FUELS

The revenue generated by WPT attracted numerous interests seeking to share in the revenue. Some of the purposes WPT revenue was earmarked for included tax credits for expenditures on energy conservation equipment, removal of excise taxes on gasohol, assistance for heating and cooling by low-income families, and subsidies for the expansion of mass transit systems. But the greatest share of WPT revenue was devoted to subsidizing the production of synthetic fuels and the extraction of oil from alternative sources such as shale. The subsidies took the form of tax credits for synthetic fuel production of $3 per barrel of oil equivalent as well as specific grants and loans at low interest for specific projects to be awarded by a quasi-governmental U.S. Synthetic Fuels Corporation.[12]

These subsidies for synthetic fuels were the consequence of the hysteria that seized the American public following the overthrow of

11. Kalt, *Oil Price Regulations*, p. 234.
12. Kalt, *Oil Price Regulations*, p. 22.

the Shah, the rapid rise in oil prices, and the shortages of gasoline that occurred in 1979. This hysteria contributed to an obsession with reducing American dependence on foreign oil sources at whatever cost, and the obsession was reinforced by a dread that conventional sources of oil were on the verge of exhaustion.

As so often is the case, the only way public officials could think of to convey an impression that they were coping with the crisis was to propose and enact new legislation and new programs and to spend money—preferably a lot. A crash program to produce synthetic fuels thus became a manifestation of a commitment to energy independence and the centerpiece of a new national energy policy.

When anyone even bothered to consider why subsidies to promote the development of synthetic fuel production were necessary, supporters would argue that private enterprise could not be relied upon to put up the capital necessary for developing a synthetic fuels industry.[13] The reasons that private enterprise would not do so usually went something like the following: (1) The technology was too advanced. (2) The amount of capital required was too large. (3) The lead times were too long. (4) The uncertainty was too great.

None of these reasons are convincing.

1. The technology of synthetic fuels is not particularly advanced, since synthetic fuels have been manufactured for over a century. The basic problem is that so far synthetic fuel production has been much more costly than obtaining energy from conventional sources.
2. The economic problem is not the amount of capital required. When the prospect of profit was clear, enormous amounts of capital were available for financing the Alaskan oil pipeline without any government financing.
3. Similarly the lead time in developing synthetic fuel projects is not unusually long. Lead times in the development of commercial aircraft are as long as or longer than those for synthetic fuel projects.
4. The uncertainties associated with synthetic fuel projects are not inherently any greater than many other projects financed by private capital.

13. The argument of the rest of this section is based on Joskow and Pindyck, "Subsidized Energy Schemes."

The only conclusion to be drawn is that the reluctance of the market to finance the development of synthetic fuels was based on a widespread opinion that such investments were dubiously profitable. The subsidies for synthetic fuels were needed to encourage private investors to take risks with taxpayers' money that they were unwilling to take with their own.

Thus while WPT discourages the exploration for and development of conventional energy sources that would otherwise have been profitable, the revenues collected from WPT are often given back to the same companies to finance projects that would not have been deemed profitable in the absence of the subsidy. The net result is that the oil companies are induced to shift their efforts from projects that promise a greater return to projects that promise a smaller return. If we assume that the decisionmakers know anything about their business, then there is likely to be a net waste of resources and a net reduction in energy output over the long term. Moreover, the WPT itself acts as a deterrent to many private investments in synthetic fuel production since it implies an increased risk that any successful synthetic fuel producer will be subjected either to price controls or to a new WPT in the future should the price of energy rise substantially once production has begun. Risk itself is no deterrent when the potential gains are large enough, but when it is made clear in advance that any exceptional gains will be confiscated by the government, even a moderate risk will appear overwhelming, unless a subsidy is provided.

WHAT CAUSES WINDFALL PROFITS

From listening to supporters of price controls or of WPT discuss windfall profits, one might conclude that windfall profits accrue to individuals or firms in a manner entirely independent of their prior behavior. The very word "windfall" suggests that the profit is entirely fortuitous and has fallen, as it were, into the lap of some lucky person who has done nothing to earn it. Thus in arguing against allowing windfall profits to accrue to U.S. producers, President Carter remarked: "[The] increase in [the] value [of American oil in existing wells] has not resulted from free market forces or from any risk taking by U.S. producers."[14]

14. Carter, *see* U.S. President, *National Energy Plan.*

Neither of the two assertions Mr. Carter made in that statement can withstand scrutiny. Although my concern here is with his assertion that the increase in value of American oil did not result from any risk taking, let me just point out the problem with his assertion that the increase did not result from free market forces. It is true that the value of oil in world markets increased because of cartellike behavior on the part of OPEC, not because of free market forces. But the cause of the oil price increase in world markets is irrelevant to the market forces that increased the value of American oil. Since oil is an internationally traded commodity, market forces require that the value of oil in America correspond to the value in world markets regardless of the specific cause of the increase in the world price of oil.

Let me now turn to Mr. Carter's assertion that the increase in the value of American oil did not result from any risk taking by American producers. If all he meant was that the cause of the increase in price had nothing to do with risk taking by American producers, then what he said was true but trivial. It would not mean that American oil producers had not taken any risks. Suppose the price of grain rises because of a bad harvest in the Soviet Union. The price increase would also be unrelated to any risk taking by American farmers. Does that mean American farmers should not capture the increase in the value of their crops?

Perhaps Mr. Carter meant that producers had borne no risk. But unless they were certain that the price of oil would rise as much as it did, it cannot be said that they were bearing no risk. If they did, in fact, know that the price would rise as much as it did, then presumably they must have made investments in developing oil-producing properties based on that conviction, in which case to deny them the return on which the investments were based is not obviously the just and proper thing to do.

However, since producers obviously could not have known the course of future oil prices in advance, they must have been bearing some risk. Perhaps they were expecting stable prices, in which case the higher prices they received were an unnecessary reward for having found the oil in the first place. Presumably this would mean that producers were undergoing no risk that prices would fall. That assumption though is impossible to reconcile with the fact that in real terms the price of oil fell in domestic as well as foreign markets between 1954 and 1971. Along with this decline in the real price of oil, there was a decline in domestic exploration and development.

By 1971 this decline had resulted in the erosion of domestic excess production capacity and in falling domestic output and reserves. Producers who had invested during this period obviously were accepting a risk of falling prices that many others were unwilling to bear.

What Mr. Carter may have meant therefore was that producers profited from appreciation in their oil reserves after they had already committed themselves to the extraction of those reserves. Once having developed producing properties, they would continue to produce even if prices turned out to be much lower than they were expected to be. If this is what he meant, then the meaning of a windfall profit to Mr. Carter must be a profit accruing to anyone who is committed, with or without the profit, to doing what one is currently doing. When someone is in a position to withdraw one's output from the market if the anticipated return is not realized, that profit, under Mr. Carter's definition, would not be a windfall. But if the investment is such that, because of very high initial start-up costs, once it is undertaken the project will be carried forward and production continued whether or not the expected price is realized, then under Mr. Carter's definition, the profit would be a windfall. But it is precisely such projects that are the most risky, and those who invest in them are bearing a greater risk than those who are not committed to continuing to produce at whatever price is realized.

EVERY PROFIT IS A WINDFALL

Economists ever since Adam Smith have been concerned with and have theorized about profits in an effort to explain their occurrence and their function. Since 1924, when the great American economist Frank Knight published a remarkable book on the subject of profit,[15] economists have come increasingly to recognize that the existence of a pure profit (that is, either income or a capital gain left over to the ownership of an enterprise in excess of what could be considered as an implicit wage for services rendered or interest on capital invested) is inconceivable in the absence of uncertainty about, and incomplete knowledge of, the future.

Every profit of course arises because someone has succeeded in buying cheap and selling dear either directly or indirectly. If you can

15. Knight, *Risk, Uncertainty, and Profit.*

buy apples for 5¢ a pound and resell them for 25¢ a pound, you will make a profit doing so if it is not very costly for you to make those transactions. Or suppose you can hire someone to work for you who is very good at knitting sweaters. Say you agree to pay that person $2 per sweater and the wool required to knit a sweater costs only $1. If you can sell the sweaters for $15 each, in effect, you will have indirectly bought the sweaters cheap and resold them dear.

What Knight pointed out was that for you to have been successful in doing either of these things, it must be that not everyone else could foresee what you were capable of doing. For if they could, they too would have tried to buy apples at 5¢ a pound and resell them for 25¢ a pound. In the process of doing so, however, they would have eliminated the price difference and the profit by driving the price up in the cheap market or driving it down in the dear one. The same would apply in the sweater case. If everyone foresaw this opportunity, either the wage of the knitter would have been bid up until the profit to you was eliminated, or if many other knitters of equal ability could be hired for the same wage, the price of sweaters would be forced down to eliminate the profit. Thus Knight showed that it is only insofar as some price-cost relationships are not generally foreseen that the lucky or the more prescient can earn profits.

The phenomenon stands out in all its clarity in the futures markets for various commodities. Those buying contracts for future delivery expect the price of the commodity to rise in the future and expect then to be able to resell the contracts at a profit. If I buy a futures contract in heating oil for delivery in December 1984 at a price of $1 per gallon, I am betting that the price of heating oil will be more than $1 per gallon in December 1984. If it is, I shall be able to take delivery for $1 per gallon and resell at the higher market price. (In practice, I would merely resell the contract at a price above $1 per gallon, something I may do at any time before the contract comes due.) Those who sell futures contracts expect the price to go down and expect then to be able to buy the contract back later at a profit. For example, if I sold a futures contract for delivery of heating oil in December 1984 at $1 per gallon, I should be betting on a price reduction that would enable me to buy heating oil at a market price less than $1 per gallon and then to sell it at the price ($1 per gallon) currently stipulated. (Again, in practice, I should merely have to buy back the contract for less than $1 per gallon at any time before the contract came due.) But if everyone knew with certainty what the future price would be, futures contracts for delivery at any given date would always have the same price, and the profits that arise from changes in the value of these contracts would never occur.

In the real world the future can never be foreseen perfectly; opinions about the future, even among experts, always differ. Those who anticipate the future most accurately and who take actions accordingly make profits, while those who anticipate the future least accurately and take actions accordingly incur losses. In a private ownership market economy, this sort of risk taking is imposed on each individual, although there are possibilities (insurance and hedging) for pooling, exchanging, and taking on offsetting risks. Let me explain further: To reduce this exposure to risk, individuals can pool them so that a lot of small and predictable losses are spread out among everyone instead of a few big losses imposed on a few individuals. Insurance companies acquire a lot of risks from individuals and then pool them. Hedging means assuming one risk that offsets another; for example, a fighter who bets that his opponent will win the fight is hedging.

In nonprivate ownership economies risks are largley collectivized, but individuals must still bear risks themselves (e.g., of goods being in short supply or of having to change jobs). Because the absence of private ownership reduces the rewards for correctly anticipating the future, precautionary or speculative behavior is more likely to be observed when private property exists than where it does not. Thus, without private ownership, hoarding in anticipation of a shortage, which if the shortage materializes is beneficial to society as well as profitable to the hoarder, cannot be rewarded. Indeed, in a fully collectivized society, it is likely to be classified as a form of criminal activity.

The point here is that any attempt to single out profits earned by a particular group of individuals as being a windfall is ultimately based on a completely arbitrary distinction. Any profit earned by anyone at any time must have some windfall element. To say that the profits earned by the oil companies were windfall profits is simply a way of saying that the oil companies reaped profits because of a decision that turned out to be lucky. But those who bought homes in the late sixties and early seventies have also earned windfall profits. Those who bought gold at $35 per ounce earned windfall profits as did those who purchased shares of Twentieth Century Fox before *Star Wars* was released. What is the basis for distinguishing between them?

Try to imagine what the world would be like if all windfall profits could somehow be confiscated. Do you think the world would be a better place? Do you think we would have more or less oil to use today if a windfall profits tax had been enforced on oil producers over the last century? For many investments it is the small probability of a large reward rather than the high probability of a modest one that stimulates entrepreneurship. If the small probability of a large reward were eliminated, much of that entrepreneurship would disappear along with it.

Perhaps you resent the fact that Reggie Jackson makes so much money for playing a child's game or that Burt Reynolds makes so much for acting in the movies. Yet it is the prospect of making such enormous salaries that induces a Jackson or a Reynolds to perform for a public that is willing to pay enough to watch them to allow their salaries to be paid.

The windfall profits that accrue to those who have previously invested in oil production are more than a gratuitous reward to those who would have done precisely the same thing even without the reward. Others could have produced oil, but they did not. The British economist Edwin Cannan used to wonder why it was that the public blamed sellers for high prices since if it were not for the sellers the price would have been higher still. The blame for higher prices, Cannan argued, belonged on everyone else who was not selling the product and hence not contributing to a lower price.[16]

Allowing those who did invest to reap the rewards of appreciating oil means that those who did, in fact, devote resources to exploring for and developing oil resources will be rewarded for ensuring that more of the increasingly precious commodity is available to the rest of us than would have been available without their efforts. If we adopt a policy of not permitting windfall gains to accrue to those who invest in finding resources that became more valuable after they have been discovered, a great part of the incentive for finding such resources will be eliminated, and we are sure to find ourselves undergoing rising resource costs in the future. The chief incentive for trying to find such resources is the expectation that they will become more valuable in the future. But if it is known beforehand that future appreciation will not accrue to those who have discovered the resource but will instead be expropriated through price controls or

16. Cannan, *An Economist's Protest*, p. 18.

taxation, how can we expect risky exploration and development to take place?

THE FAIRNESS ARGUMENT

Despite the inefficiencies created by WPT, you might say that it is not fair for oil producers to earn such enormous profits – particularly when those profits arose from circumstances that imposed severe hardships on nearly everyone else. Those who profit from the misfortune of others are commonly regarded with reprehension and scorn. To avoid reprobation, one must demonstrate that the profit was the result of some evident merit, effort, or sacrifice on one's own part.

Having very little demonstrable merit and with no particular effort or sacrifice that could be associated with the rising world price of oil, the oil companies seemed to lack any vestige of moral entitlement to the increase in the value of their property. Arbitrary wealth transfers effected by blind market forces are rarely accepted with the same philosophical spirit by those who lose from them as by those who gain. Such transfers therefore frequently evoke, in the name of fairness, countermeasures designed to preserve the prior distribution of wealth.

Just how fairness requires the restoration of the prior distribution of wealth is, however, not quite clear since the prior distribution was the result of the same sorts of blind market forces that brought about the wealth transfer at issue. It is true that in the case of the increase in oil prices the transfer was brought about by actions that would be illegal if they had occurred within the United States. But from the point of view of the American market, OPEC was an exogenous factor to which domestic market forces had to adjust.[17] In what sense, then, can the prior distribution of wealth be judged to have been more fair than the later one?

One might argue that income and wealth differences are, in themselves, unfair and ought to be mitigated insofar as possible. This is an honorable and widely held – though not unchallenged – moral posi-

17. This does not mean that the domestic response was irrelevant to OPEC's position. On the contrary, I already have shown how the response actually assisted OPEC because of controls and EP. But the fact that OPEC was behaving as a cartel was not something that made a qualitative difference in the way that domestic market forces operated.

tion. Of course few would advocate full equality as an ideal because of the manifest reduction in everyone's income that would result from attempting completely to equalize incomes. Thus the most influential recent formulation of the principle has argued that income differences may be tolerated only insofar as they lead to an improvement in the position of the least well-off groups within society.[18] Whatever its merits, this principle clearly has nothing to do with a policy that would single out for special punitive treatment income or wealth derived from one particular source or by one particular group. Reducing differences in income and wealth means taking from those who have most and giving to those who have least. It does not mean taking from oil producers and giving to oil consumers or to oil refiners or to producers of synthetic fuels. No appeal to fairness as a principle can justify these arbitrary redistributions.

Any legitimate concern with inequalities arising from the increase in oil prices could have been addressed by providing aid to low-income individuals and families adversely affected by increased oil prices. Nor should it be forgotten that the moral argument that sanctions redistribution to the poor is entirely indifferent as to whether the recipient's poverty is the result of increased oil prices or, say, the low wages that are paid to agricultural workers in the sunbelt. Those harmed by increased oil prices do not acquire a claim to greater concern than those less affected by increased oil prices but who nevertheless live in greater poverty.

Still you might say that a windfall profit is essentially an unmerited gain to which private individuals have no moral claim. But few of us would be ready to accept the implications of such a moral stance if applied consistently. Each day in the stock market and in the commodity exchanges billions of dollars in profits (and losses) are made. As I have argued above, all such profits are essentially windfalls. Are we then to regard all these profits as unmerited windfalls and liable to partial or (if we accept the logic of the principle) total confiscation? To do so would destroy or drive underground not only the stock market and the commodity markets but also just about every asset market one can imagine.

In sum, it is difficult to formulate any principled moral position that would justify discriminatory treatment of income earned by a particular group of people or by those earning income from a par-

18. Rawls, *Theory of Justice*.

ticular source. Nevertheless, something might be said in favor of discriminatory treatment if such treatment promoted greater efficiency in the use of resources. But we have already seen that the discriminatory treatment has precisely the opposite effect. It not only makes the use of resources less efficient, it also runs directly counter to the avowed goal of promoting the discovery and development of additional energy sources.

CONCLUSION

WPT continued a number of unfortunate trends in policy toward the oil industry. Perhaps the oldest of these trends is transferring wealth between different sectors of the oil industry. Under WPT these are reflected in the special treatment given to independent producers and small refiners as well as in subsidies awarded to producers of synthetic fuels. A more recent trend is to penalize the production of oil and to attempt a kind of discriminatory taxation in which "old" oil is taxed at higher rates than "new" oil; the theory being that old oil will be produced no matter what, so it can be taxed at virtually any rate without affecting supply, while lower tax rates on new oil will induce further exploration and development. Although there is undoubtedly some truth in the theory, it fails to give potential oil producers credit for having enough intelligence to figure out that all new oil will eventually become old oil and then become ripe for similar exploitation. Discriminatory taxation of old oil must therefore also reduce the amount of exploration for and development of new oil that takes place.

Both of these trends in policy toward the oil industry have also been evident in the control of natural gas prices. The controls over natural gas prices and their effects are the subject of the next chapter.

Chapter 10

NATURAL GAS
Will We Learn From Our Mistakes or Repeat Them?

INTRODUCTION

In previous chapters I have recounted the destructive effects of price controls and other interventions in the markets for crude oil and refined products. In almost every instance these effects have been contrary to what most, though by no means all, supporters of such measures had intended. The principle of course is an old one, but even though Adam Smith masterfully explained two hundred years ago that controls over the marketplace always have undesigned, unintended, and unwanted effects that frustrate the intentions of those supporting the controls, the same lesson it seems must be relearned from experience by each generation.

If the decontrol of the prices of crude oil and refined product prices is any indication, some learning indeed seems to have taken place in recent years. Nevertheless, WPT remains in effect on crude oil prices as do price controls on most natural gas at the wellhead. And even though the results of the oil price decontrol contradicted the dire predictions opponents of decontrol had made, similar predictions are still taken seriously by policymakers and the public.

What I want to do in this chapter therefore is to discuss the market for natural gas and how controls have affected it. The general character of those effects should not be too difficult to surmise, but the special features of the natural gas market (which often give rise to spurious arguments suggesting that controls are necessary for the market to function fairly and efficiently) and the continuing debate about whether to maintain existing controls over natural gas prices make it advisable to spell out the effects of controls in some detail.

In order to understand those effects, it will be helpful first to go over some basic facts about natural gas—where it is found, how it is transported, and how it is used—and then to review briefly the historical development of the natural gas market and particularly the development of regulatory control over that market.

THE CHARACTERISTICS OF NATURAL GAS

A gaseous mixture of hydrocarbons—primarily methane—natural gas is found in geological formations similar to those containing crude oil. About 30 percent of all gas reserves are associated with oil, but the remainder are not associated with oil.[1] When associated with oil, gas is extracted along with oil since the gas will escape from any open well to relieve pressure beneath the surface. If there is no means for gathering, transporting, and marketing the gas, it must be flared at the well. Before the technology was developed to gather, transport, and market gas, it was basically a nuisance associated with the production of oil. Any wells from which only gas could be extracted were simply capped.

For a long time after crude oil began to be produced in substantial quantities, the commercial use of gas remained quite limited. Only in the Appalachian region, where large amounts of gas were found in the late nineteenth century, was gas produced commercially on a significant scale. The primitive pipeline technology then extant sufficed to transport gas from the producing fields in the region to the major markets of Buffalo, Cleveland, and Pittsburgh and to the many smaller towns in the vicinity.

It was only with the development of seamless welded pipes in the twenties that the gas industry began to grow in other parts of the country. During the thirties over 1,000 miles of pipeline were built in the United States. By 1950 more than 100,000 miles of pipeline had been built, and today there are more than 260,000 miles of gas pipeline in the United States.[2]

1. Murphy, O'Neill, and Rodekohr, "Natural Gas Markets," p. i.
2. Murphy, O'Neill, and Rodekohr, "Natural Gas Markets," p. i.

Gas is produced by numerous companies, ranging from small independent firms operating only a few wells to the major oil companies that produce a substantial–though not overwhelming–share of the nation's gas output. Most gas is sold at the field to pipeline companies that transport the gas from the field to the various markets, which are often more than 1000 miles away. Producers sometimes sell gas to major industrial users or utility companies. Some pipelines own producing gas wells, but these account for only a small share of total output. The pipelines sell gas at the local market to local gas companies and to major industrial users. The local gas companies then supply individual households as well as commercial and industrial users of gas.

The price at the wellhead of most gas is subject to federal price ceilings. Rates charged by interstate pipelines are also subject to federal price regulations. Public utility regulation by state and local governments controls the rates set by local gas companies.

In 1981 total gas consumption in the United States was about 20 trillion cubic feet (tcf.). Gas consumption in the United States reached its peak in 1972 when 22.1 tcf. were consumed.[3] About 40 percent of all the gas used in the United States is consumed by industrial users. Of this amount, about 64 percent is consumed by the chemical, petroleum refining, and primary metals industries. Residential use accounts for about 25 percent of the nation's consumption of natural gas. Households use natural gas for space and water heating and for cooking. Commercial establishments, which also use gas primarily for these purposes, account for about 12 percent of the nation's gas consumption. Electric utilities use natural gas to run power plants and consume about 15 percent of the gas sold in the United States.[4]

Natural gas consumption follows a seasonal pattern. Consumption during the winter reaches a peak that is nearly double the amount consumed in the summer. The heavy demand during the winter is due to the increased residential and commercial demand for gas to provide space heating. The cyclical demands by residential and commercial users are partially offset by the countercyclical demands of industrial users. Since gas must be delivered to consumers by pipeline, the capacity of the pipeline (a function of the diameter of the

3. Murphy, O'Neill, and Rodekohr, "Natural Gas Markets," p. ii.
4. Murphy, O'Neill, and Rodekohr, "Natural Gas Markets," p. ii–iii.

pipe and the pressure within the pipeline) must suffice to handle the peak demand of users serviced by the pipeline.

In order to reduce the cost of adding capacity to accommodate peak demand (either by using larger pipe or by adding pumping stations to increase pressure), pipelines seek to discourage demand during peak periods. They do so by offering discounts to users that contract for service that may be interrupted during peak demand periods when demand would otherwise exceed capacity. Generally, it is the industrial users that can switch to alternative fuels like residual fuel oil or coal relatively cheaply and electric utilities, which normally face slack demand in the winter, that avail themselves of the discounts offered on interruptible contracts.

Another method of economizing on capacity to handle peak demand periods has been for pipelines to store gas in the vicinity of their markets for later use. The variation in inventories of gas held allows production to remain fairly stable throughout the year, while accommodating the seasonal fluctuations in demand that exist despite the incentives for agreeing to interruptible service during peak demand periods.

THE NATURAL GAS MARKET BEFORE 1938

Large quantities of natural gas first began to be produced and marketed around the turn of the century in the Appalachians.[5] Areas of western New York and Pennsylvania, West Virginia, and eastern Ohio were supplied by wells concentrated largely in West Virginia, western Pennsylvania, and southeastern Ohio. Most of the distribution was controlled by the Standard Oil Company of New Jersey (Exxon) and the Columbus Gas and Electric Corporation.

Gas production and consumption in the Appalachians expanded rapidly until 1917 when gas production reached its peak of .519 tcf. The following year production dropped sharply to .459 tcf.[6] The abrupt reversal of rapid growth in production and consumption gave rise to fears of ultimate exhaustion of reserves and the disappearance of the market for natural gas in the region.

5. The account in this and the following two sections draws heavily on Kitch, "Regulation for Natural Gas."

6. Kitch, "Regulation for Natural Gas," p. 249.

Markets reacted predictably—prices rose. In Pennsylvania and Ohio gas prices doubled between 1916 and 1922.[7] As a result of higher prices, many industrial users of gas switched to alternative fuels so that the percentage of gas in the region consumed by industrial users declined from 64 percent in 1916 to 44 percent in 1925.[8] The increase in price also stabilized the production of gas in the area. From 1925 to 1965 production hovered about .33 tcf per year.[9]

Parenthetically, I would point out that recent improvements in exploration technology along with rising prices for new gas have stimulated interest over the past few years in the Appalachians as a potentially significant source of new gas supplies. An overthrust belt in the Appalachians geologically similar to the very productive overthrust belt that is being developed in the Rocky Mountain area is now expected to turn the Appalachians once again into a major source of natural gas.

Before passage of the Natural Gas Act of 1938, there was no regulation of gas prices at the federal level, but such attempts were made in individual states. For example, in West Virginia, the leading producing state in the region after 1909, prices were under the control of the state public utility commission, which applied traditional cost-of-service criteria in setting rates. In the winter of 1916–17 there were prolonged interruptions of the interruptible supplies that severely disrupted industrial activity in the state. What happened was that the neighboring states of Pennsylvania and Ohio allowed higher prices than West Virginia did, thereby inducing producers in West Virginia to sell in the interstate market instead of locally.

Responding to the increasing diversion of supplies to the interstate market, the West Virginia legislature passed a law requiring West Virginia gas producers to satisfy all the demands of their West Virginia customers before exporting any gas out of state. Pennsylvania and Ohio promptly filed suit challenging the constitutionality of the statute. In 1923 the U.S. Supreme Court held that the statute did violate the interstate commerce clause of the Constitution.[10]

Regulators in states that used natural gas were equally powerless since the only alternative to allowing higher prices was for service to

7. Kitch, "Regulation for Natural Gas," p. 249.
8. Kitch, "Regulation for Natural Gas," p. 249.
9. Kitch, "Regulation for Natural Gas," p. 249.
10. *Pennsylvania* v. *West Virginia*, 272 U.S. 544 (1923).

be abandoned. It became clear that regulation of the gas market would have to be undertaken at the federal level to be effective.

Development of the gas fields in the Midwest proceeded more slowly. In the early 1900s a local network of pipelines in eastern Kansas and Oklahoma distributed gas near the oil fields in which gas was also being produced. Because of fears that the gas would soon be exhausted, Oklahoma attempted to prohibit the export of gas out of state. This attempt was struck down by the U.S. Supreme Court.[11]

The early fears that the supply of gas would be depleted were soon dispelled as new gas fields continued to be discovered and more pipelines built. In the 1920s two huge fields containing enormous quantities of nonassociated gas were found. One was in southwestern Kansas and the other in the Texas Panhandle. Abundance, however, created problems of its own. Pipelines did not have enough capacity to transport all the available gas. Associated gas was flared at the well and wasted if there was no market for it. Although nonassociated gas could be left in the ground, the kind of common property resource problem that we encountered in Chapter 7 in connection with oil fields with several independent property interests was even more pronounced in the case of gas. There was typically a rush by all the property interests in a gas field to sell gas, even at very depressed prices, lest whatever remained beneath a given property be extracted from a well on adjoining property.

The fears and difficulties occasioned by the opposite problems of short supply and abundance prompted Congress to request that the Federal Trade Commission (FTC) conduct a study of the natural gas industry. The FTC report published in 1935 had a major influence on the Natural Gas Act (NGA) of 1938.

THE NATURAL GAS ACT OF 1938

The FTC report dealt primarily with the problems in the Appalachian fields, leaving the problem of overproduction to individual states, many of which had already taken steps to deal with this problem in relation to oil. The role of monopolistic pipelines that were supposedly free to raise prices without regard to the consumer interest

11. *Oklahoma* v. *Kansas Nat. Gas. Co.*, 221 229 U.S. (1911).

was the principal concern of the FTC. It recommended that interstate pipeline operations be subjected to traditional utility rate regulation by the federal government. The FTC report did not find that gas producers were exercising any monopoly power. Indeed the report cited the treatment of independent producers as an example of the monopolistic abuses in which the pipeline owners had engaged.[12]

The charge against the pipeline owners was that, because of their limited numbers and the difficulty of switching from one pipeline to another, once a producer or a customer had been linked to a pipeline, the pipeline acquired monopolistic control over its customers and monopsonistic control over its suppliers.[13] If pipelines could exercise monopoly power against consumers and monopsony power against producers, there would be a strong case for regulating them in order to protect both suppliers and customers.

In 1938 Congress responded to the FTC report by passing NGA. NGA charged the Federal Power Commission (FPC) with responsibility for controlling the rates set by interstate pipelines. But a critical ambiguity lurked in the language of the act. On the one hand, NGA was supposed to "apply to the sale in interstate commerce of natural gas for resale," which would seem to have included sales at the wellhead to interstate pipelines. On the other hand, it also stated immediately afterwards that it was not to "apply to the production or gathering of natural gas," which seemed to exclude all sales at the wellhead, even those to interstate pipelines.[14]

With rare bureaucratic self-restraint, FPC construed the act to mean that sales by independent producers were indeed exempt from the jurisdiction of NGA. The industry, fearing a later reversal of this interpretation, sought an explicit Congressional declaration that producers were to be exempt from price controls. Congress did pass an amendment to NGA to this effect, but President Truman vetoed the bill, citing as a reason for not foreclosing the option of extending regulation to producers the possibility that dwindling supplies might in the future enable producers to raise prices.[15]

12. *Final Report of the FTC on Utility Corporations*, S. Doc. 92, Pt. 84-A, 70th Cong., 1st Sess. 101-10 (1936) pp. 184-98, 608.

13. A monopsonist is a buyer (strictly speaking the only buyer in the market) that can reduce the price it pays its suppliers by reducing the amount it purchases. A monopsonist is thus the analogue to a monopolist on the demand side of the market.

14. 15 U.S.C. Sec 717(b) (1964).

15. 1950 *Public Papers of the Presidents of the United States*, p. 257.

The issue was not finally settled until 1954 when the Supreme Court decided the case of *Phillips Petroleum Co.* v. *Wisconsin.*[16] The Court ruled that the language of NGA did indeed apply to the prices charged by producers at the wellhead. The exclusion of the "production and gathering of natural gas" from the provisions of NGA was taken by the court to refer only to the installation and operation of facilities used in production and gathering.

Responding to the *Phillips* decision, independent producers again sought an unequivocal expression of Congressional intent that producers were not to be subject to price regulation. With a Republican President in office, it was expected that a bill restoring the exemption would be signed if passed by Congress. Such a bill was indeed passed by Congress. Although supporting the bill on its merits, President Eisenhower nevertheless felt obliged to veto the bill because improper lobbying techniques (allegedly including direct payments to a senator in return for a favorable vote on the bill) had been used to secure support for the bill in Congress.[17] The *Phillips* decision thus remained the controlling interpretation of NGA, and FPC was obliged to regulate the terms of the sale of natural gas at the wellhead.

WARTIME SHORTAGES OF NATURAL GAS

Experience about the potentially disastrous effects of controls on natural gas prices was not lacking at the time the Supreme Court decided to extend controls to producers. Natural gas, along with all other commodities, had been subject to controls during World War II. By the time of the war, new pipelines had connected the massive Hugoton field in Kansas and the Panhandle field in Texas with major consuming markets in the Midwest. A number of large midwestern cities had become dependent on natural gas for heating and electricity. Furthermore, because of heating oil shortages that had already arisen, many additional users were switching to natural gas.

The price of natural gas was frozen as part of a general policy enforced by the Office of Price Administration. Rising incomes, increasing industrial production, and fuel switching by consumers

16. 347 U.S. 672 (1954).
17. *New York Times*, 7 February 1956, pp. 1, 22.

unable to obtain heating oil began to create serious shortages in 1944. Interruptible users found themselves cut off from supplies for extended periods of time, and shutdowns of war-related production resulted. Shortages were most severe in the Appalachian region but were also evident in the mid-continent because shortages of materials prevented pipelines from being built fast enough to keep up with the growing demand.

The winter of 1944, the coldest in twenty-five years, was marked by three periods of emergency shortage during January and February. These shortages were so severe that a number of cities in Ohio, including Cincinnati, Cleveland, Columbus, and Dayton, were threatened with a complete loss of service. Had a loss of service occurred, safety procedures required to avoid any explosions when service was restored could have postponed the resumption of service for as much as two months.

During the second emergency period, total curtailment of gas service reached 55 percent. The curtailment of gas service to critical war factories was 24 percent.[18] In order to avoid complete curtailment of the Appalachian system, the Michigan Consolidated Gas Company—which supplied Detroit, a major center of war industry—instituted complete curtailment to allow diversion to the Appalachian system to prevent a loss of service there. Fortunately complete curtailment in Detroit lasted only half a day. Nevertheless, Army ordnance estimated that 300,000 tons of steel for the war program were lost because of the gas curtailments in addition to substantial amounts of finished products.[19]

By the time of the *Phillips* decision, however, the wartime experience had been forgotten. And it was to be almost twenty years before the price controls on gas at the wellhead required by *Phillips* gave rise to shortages comparable to those of World War II.

THE NATURAL GAS MARKET AFTER THE *PHILLIPS* DECISION

The production and consumption of natural gas in the United States expanded rapidly after World War II. Pipelines connected both the

18. Frey and Ide, *Petroleum Administration*, pp. 231–32.
19. Frey and Ide, *Petroleum Administration*, pp. 231–32.

east and west coasts with the major gas producing fields in the Gulf Coast and the Southwest. Nor, for a long time, was this expansion much restrained by the *Phillips* decision and the beginnings of federal regulation of the field market for natural gas.

Two aspects of FPC regulation in the period immediately following *Phillips* should be noted.

1. At first FPC attempted to regulate producers by subjecting each producer to price ceilings based on the producer's own cost of service in the manner of traditional public utility regulation. Since such regulation involves setting rates based on the cost of service incurred by one or at most a very few firms, it is easy to imagine how hopeless the task of determining permissible price limits for thousands of individual gas producers must have been.
2. FPC did not undertake to review the prices of contracts already in force. What the Commission required was that all increases in wellhead prices be submitted for approval. By 1960 over 2,900 applications for increased rates had been submitted to FPC for approval. At year's end only 10 of these requests had been acted upon.[20]

Sheer administrative overload obviously required that the Commission seek a different approach to regulating the wellhead prices of natural gas, and further Supreme Court decisions forced FPC to adopt additional changes in its regulatory approach. In a decision rendered in the case of *Atlantic Refining Company* v. *Public Service Commission* (referred to in the literature as *CATCO*) in 1959,[21] the Supreme Court ruled that FPC was obligated to review the initial terms of a contract to determine whether the proposed price was "in keeping with the public interest." In subsequent decisions, the Court evolved the doctrine that FPC was required to ensure that the price of gas in new contracts was "in line" with already existing prices for gas under similar conditions. In effect, the Court required that FPC impose a virtual freeze on natural gas prices in new contracts.

In an attempt to break the administrative backlog of cases, the FPC announced in 1960 that it would regulate gas prices on an area basis from then on.[22] FPC divided the country into five producing

20. Breyer and MacAvoy, "Regulating Natural Gas," p. 174.
21. 360 U.S. 378 (1959).
22. Phillips Petroleum Co., 24 FPC 537, 542–48 (1960).

areas and soon began proceedings to determine price ceilings for each area. In the meantime gas prices were essentially frozen at their 1959–60 levels.

Before going on with the development of regulatory control over natural gas prices at the wellhead, we should briefly consider the reasons that were advanced to justify this control. Two conceptually distinct, but frequently confused, reasons emerge. First, it has been suggested that controls were necessary because of monopoly power exercised by the producers. Second, it is said that controls prevent windfall profits from accruing to gas producers as gas becomes more scarce.

The argument about monopoly power asserts that given the limited number of major gas producers, the market at the wellhead could not be competitive. Thus the monopolistic prices set at the wellhead would perforce be passed on to consumers despite the regulation of pipelines. The monopoly argument was reinforced by the belief that producers were somehow in a superior bargaining position with respect to pipelines. Having invested in costly pipeline equipment to transport the gas from field to market, pipeline companies supposedly could not afford any interruption in gas deliveries from the producer and would, unless protected by price controls, pay any price demanded by the producer.[23] The latter argument is clearly faulty since it ignores the corresponding costs that a producer would incur if a pipeline company were unwilling to buy gas after the pipeline had been installed. The bargaining positions of pipeline owners and gas producers are, in truth, symmetrical.

Be that as it may, the Supreme Court was obviously influenced by a belief or at any rate a fear that producers were indeed exercising monopoly power. The Court made reference in *CATCO* to rising field prices during the fifties and concluded that this trend was an indication of monopoly power.

Subsequent research, however, has shown that the increasing prices for natural gas at the wellhead were caused by the declining monopsony power of pipeline companies as the market for natural gas expanded nationally during the fifties.[24] Producers in areas that had been serviced by only one company began to reap the benefits

23. Douglas, "Consumer of Natural Gas."

24. See Kitch, "Regulation for Natural Gas," pp. 262–64 and MacAvoy, *Price Formation.*

of competition for their gas by new pipeline companies that were supplying the rapidly increasing demands of their customers.

Price increases came primarily in fields with new pipelines. Rising prices were also associated with increasing price differences among producers to reflect such factors as distance, volume, and length of contract, all of which would be more influential under competitive market conditions than under monopsonistic conditions.[25] Furthermore the evidence does not show that higher prices in the field were passed along to consumers as they would presumably have been if higher field prices had been the result of monopolistic restrictions by the producers.[26]

Nor does the evidence on the numbers of and market shares of producers support the charge that they exercised monopoly power. Market shares of producers of natural gas were, and continue to be, low by any standard of comparison. In the early 1960s, for example, the four largest gas producers controlled less than 10 percent of the national production, and the fifteen largest, less than 50 percent. Even when markets are defined regionally, concentration ratios remained quite low. In the Permian Basin of western Texas and eastern New Mexico, the five largest producers accounted for less than 50 percent of production.[27]

The second reason for controlling natural gas prices was to limit the economic rents that might accrue to low-cost producers of gas if increased demand required production from high-cost sources. In the absence of regulation, the market price would rise to induce production from high-cost sources with the result that the low-cost producers would realize substantial windfalls. The difficulty with regulation designed to eliminate such rents or windfalls is that imposing a price low enough to prevent windfalls also discourages the production from high-cost sources required to clear the market. Moreover, maintaining a low price encourages consumption and exacerbates the shortage.

One response to the problem of discouraging production from high-cost sources by price control is to control the price of low-cost intramarginal sources while allowing higher prices for the marginal high-cost sources. But if this is done without controlling the prices

25. MacAvoy, *Price Formation.*
26. Kitch, "Regulation for Natural Gas," pp. 262-64.
27. Breyer and MacAvoy, "Regulating Natural Gas," p. 166.

that pipelines may charge consumers, the controls on prices at the wellhead simply transfer rents from producers to the pipeline companies. And if the prices pipeline companies charge are controlled (which they are), then the transfer from producers to consumers amounts to a subsidy to consumers for consumption that means either (1) the shortage will persist if a limit is placed on the price of gas from marginal sources, or (2) the price of gas from marginal sources will be driven even higher in order to elicit the supplies demanded by subsidized consumers. (See Appendix B.)

Both of the above reasons seem to have played a role in motivating the controls on natural gas, though neither the courts nor FPC attempted to distinguish conceptually between them. Nevertheless one can detect a gradual shift in emphasis toward the latter as a ground for regulation. This shift can be seen in the concern expressed by the Commission in some opinions that low prices might cause inadequate supplies in the future.[28] Such a fear is out of place if a monopoly is made to reduce its price. The reduced price in fact elicits an increased supply from the monopolist in order to meet the greater amount demanded by the public at the lower price. The conceptual shift is also evident in the adoption by the Commission of different ceilings for old gas and new gas in its *Permian Basin* opinion.[29] I shall discuss the implications of this decision later.

Let me now return to the freeze that was imposed on natural gas prices at the wellhead in consequence of the various Supreme Court decisions. What effect did the freeze have? As usual one must distinguish between the short-run and the long-run effects of the freeze. In the short-run the average price of natural gas at the wellhead, which had been rising, leveled off almost immediately. This would seem to indicate that the freeze achieved its immediate objective of preventing further increases in the price of natural gas. But in fact another conclusion emerges. The average price of natural gas is calculated not only from the price of gas in new contracts, but also from the price of gas in old contracts. Since the percentage of old gas is continually diminishing, the average price would still have continued to rise as the share of old gas declined, even if the price of new gas had been frozen at 1959–60 levels. That the average price of gas stabilized almost immediately suggests that the price of gas in new

28. Breyer and MacAvoy, "Regulating Natural Gas," p. 171.
29. Area Rate Proceeding 61-1 (Permian Basin), 34 F.P.C. 159 (1965).

contracts was actually falling somewhat below levels permitted under the freeze.[30]

If one insists on imposing a ceiling on the price of some product, it is always wise to do so when the price has reached its peak. That way one is assured of the credit for stable or falling prices without also having to deal with, or take the blame for, the shortages to which a binding ceiling necessarily gives rise. The trick is to know when the price has peaked and also when the price will begin to rise again so that the price ceiling can be relaxed before a shortage is manifest. But anyone who could actually perform the trick would not waste time trying to control prices, but would undoubtedly devote these skills to becoming rich by speculating in the futures market.

Imposition of controls on gas prices at the wellhead still left part of the market for gas unregulated since intrastate sales were explicitly excluded by the language of NGA. Thus an unregulated intrastate market developed in addition to the regulated interstate market. A comparison of prices in the interstate and the intrastate markets confirms that prices in the interstate market were not, at first, bound by the freeze. If the freeze had been binding, one would have expected prices in the uncontrolled intrastate market to be consistently higher than the prices in the interstate market as, indeed, they were during the seventies when controls on prices in the interstate market were binding. But in the early sixties prices in the intrastate market were consistently below those in the interstate market.

This price differential seems to have been the result of the regulatory costs—for example, the cost of administration, the cost of delay, and the cost of anticipated future price freezes—that were imposed on the interstate market but were not incurred in the intrastate market.[31] These costs were ultimately passed along to consumers in the interstate market in order to make selling in the interstate market as attractive (prospectively, net of regulatory costs) as selling in the intrastate market.

30. Kitch, "Regulation for Natural Gas," pp. 265-69.
31. See Gerwig, "Natural Gas Production," pp. 69-92.

THE DEVELOPMENT OF SHORTAGES IN THE INTERSTATE MARKET

During the early sixties the price of gas sold in new contracts seems to have remained below the ceilings imposed by FPC. Although the price ceilings were not binding in the early sixties, and reserves continued to expand until 1967, this does not mean the ceilings were not discouraging exploration, development, and production. The kinds of regulatory costs just mentioned, particularly the expectation that future price increases would not accrue to owners of reserves, must have inhibited the search for new reserves. As early as 1964 the ratio of new reserves to increased production was insufficient to provide the fourteen-and-a-half-year inventory of reserves that purchasers of natural gas seem to have been demanding during the fifties.[32] The ratio of reserves to production, that is, the number of years of production at current rates that could be sustained by the available reserves, is an indication of the security or insurance against possible future interruptions that was being provided to purchasers of natural gas. Another indication of potential shortages looming in the future was that by 1966 the price of gas sold in new contracts in the intrastate market surpassed the price of gas sold in new contracts in the interstate market.[33]

Despite the warning signals, FPC took no action to ease the price controls that were leading to future shortages. Late in 1960 FPC did announce its intention to switch from a producer's cost-of-service method of setting price ceilings for gas prices to an area-rate method in order to manage an administrative task that had become impossible. But in the following decade FPC was only able to conclude hearings and deliberations on price ceilings in two of the five gas producing areas it had demarcated. Prices in the remaining areas remained essentially frozen at 1959–60 levels with no adjustment for inflation.[34]

In the most important decision on price ceilings in which FPC rendered its 1960 policy change, the *Permian Basin* opinion, it adopted a two-tier system of ceiling prices. A lower ceiling price of

32. Breyer and MacAvoy, "Regulating Natural Gas," p. 181.
33. Bupp and Schuller, "Natural Gas," p. 73.
34. Breyer and MacAvoy, "Regulating Natural Gas," pp. 163–64.

14.5¢ per mcf. was imposed on all gas committed to interstate commerce before January 1, 1961 and for all gas associated with crude oil. A higher ceiling price of 16.5¢ per mcf. was set for new nonassociated gas committed to interstate commerce after January 1, 1961.[35]

FPC's reasoning was that the supply of old gas and the supply of associated gas were less sensitive to price than was the supply of new gas. It concluded that a higher ceiling price was not necessary for the former categories since little additional gas would be forthcoming from those categories at a higher price. FPC further reasoned that the supply of new nonassociated gas was sensitive to price. It thus allowed the higher ceiling price for new nonassociated gas to reflect the added costs of finding and producing new supplies.

To determine the ceiling price for new gas, FPC sought to estimate the cost of developing recent gas reserves in the Permian Basin. It did so by calculating the historical costs incurred by gas producers in the Permian Basin in order to develop reserves since 1960. This procedure is another instance of the pernicious influence of the cost-of-production fallacy that I discussed in Chapter 8. It assumes that cost can be measured and calculated independently of the price and that once cost has been measured an appropriate price based on that cost can be determined.

Of course, it is obvious once one begins to think about it that the costs incurred by gas producers must in fact have been determined by the price ceiling under which they were operating. If a price of 14.5¢ per mcf. were allowed, no producer would willingly have incurred costs in excess of 14.5¢ per mcf. to develop new reserves. Indeed, if a ceiling of 1¢ per mcf. had been in effect, the Commission would probably have concluded that the cost of finding new reserves was not more than 1¢ per mcf. At most, such estimates could have determined how well gas producers had anticipated the reserves they would obtain from their expenditures on exploration and development.[36]

Thus the procedure FPC adopted for calculating the "new" price under its two-tier price system virtually guaranteed that the incentive for finding new reserves would be inadequate. This made future shortages inevitable if the policy were maintained for long. Another

35. Kitch, "Regulation for Natural Gas," p. 277.
36. See Breyer and MacAvoy, "Regulating Natural Gas," pp. 175-76.

aspect of the two-tier price system made shortages even more likely. The rationale for a two-tier system is to prevent rents or windfalls from accruing to producers of old, and presumably low-cost, gas. If the rents are to accrue to consumers instead of to the pipeline companies that deliver the gas, the price of gas to consumers must be the weighted average of the low-priced and the high-priced supplies the pipelines provide to consumers. But this necessarily involves a subsidy to the consumption of gas since, under this arrangement, consumers pay less than the marginal or replacement cost of the gas they are consuming. (See Appendix B.)

The two-tier price system implemented by FPC in its *Permian Basin* decision did not permit producers to recover their marginal costs. The gap between production and demand could be covered temporarily by drawing down reserves, but eventually the latent shortage had to manifest itself.

Beginning in 1968 total gas reserves in the United States (excluding Alaska whose gas reserves are not accessible to the lower forty-eight states or Hawaii) declined each year until 1980. And because of a growing price disparity between the interstate and the intrastate markets, whatever new reserves were being found went almost exclusively into the intrastate market. In 1969 the difference between the intrastate and interstate prices for gas sold in new contracts was 6¢ per mcf.; in 1971 it was 18¢ per mcf.; and in 1975 it was 83¢ per mcf.[37] From 1969 to 1976 additions (including revisions) to reserves committed to the interstate market in the lower forty-eight states oscillated around zero.[38] Despite these signs of shortage, FPC made no effort to raise the ceilings on interstate gas prices until as late as 1974. Its belated efforts then were far from sufficient to solve the problem that had been developing for two decades.

Nor is it surprising that FPC acted in a way that would inevitably result in future shortages. The shortages, after all, would not be perceived immediately while the higher prices that could have prevented their occurrence would be. Had they permitted higher prices, the commissioners could have expected little credit for the failure of shortages to materialize years later. But they would, undoubtedly have been charged with having betrayed consumers and sold out to the gas industry. When the shortages did occur because of their fail-

37. O'Neill, "Natural Gas Markets," pp. i–ii.
38. Bupp and Schuller, "Natural Gas," p. 73.

ure to raise ceiling prices, commissioners could easily shift the blame for the shortages from themselves to the industry.

Even though gas production continued to increase until 1973, the latent shortage of the late sixties and early seventies finally became visible in the winter of 1972-73. During that winter, curtailments of gas deliveries, even to noninterruptible customers, became quite common. With heating oil also in short supply that winter—owing to the controls on its prices—there was little opportunity for gas users to switch to heating oil. The attempt by gas users in the South to switch to propane, which is the closest substitute for natural gas, resulted in severe shortages of propane since its price was also under control.[39]

It was not until June 1974 that FPC could bring itself to revise its price ceilings on natural gas. FPC dropped its area-rate policy and adopted a uniform national ceiling of 42¢ per mcf. on new natural gas sold after January 1, 1973. In December it raised the ceiling to 50¢ per mcf. and allowed the ceiling to increase by 1 cent per mcf. per year beginning in January 1975.

The increase in the ceiling on new gas prices was too little and too late to have much effect on the shortages. Despite the price increase in the interstate market, the gap between the price of gas in interstate and intrastate markets continued to grow. Each winter from 1972-73 to 1978-79 was marked by widespread shortages that resulted in significant curtailments of service in colder parts of the nation.

To deal with the curtailments, priorities were established among different categories of users to determine the order in which users would be cut off. The priorities were: (1) residential users, (2) industries using gas as feedstocks, (3) industries using gas boilers, (4) electric utilities. Lowest priority users were to be cut off first. The inequity and inefficiency of such a system of priorities become evident once one realizes that the priorities presume in effect that the least important residential use is to be preferred over the most important industrial use. The priorities are probably based on some evaluation of the importance of the various uses of gas undertaken by the different categories and the difficulty for users in each category to find substitute fuels for those uses. But establishing a system of priorities sacrifices the most important uses in a lower priority for the least im-

39. Bupp and Schuller, "Natural Gas," Chapter 5.

portant uses in a higher priority. This necessarily results not only in waste but in gross waste.

It is no small virtue of the price system that all users are given the opportunity (or are subjected to the obligation) of choosing which of their uses of any resource are worth maintaining and which must be sacrificed. Because such choices are marginal and not categorical, the total sacrifice will be minimized.

One consequence of the shortages was to encourage the movement of major industries from the colder regions of the country, where curtailments of natural gas deliveries to industry were common during the winters, to the sunbelt states, especially those like Texas, Louisiana, and Oklahoma that had active intrastate natural gas markets in which curtailments were not a problem. Thus the attempt of the large industrial states to continue consuming natural gas produced in other states at less than market prices was ultimately self-defeating because it led to a departure of major industries in search of more secure, even if higher-priced, supplies of gas.

Faced with worsening shortages in November 1976, FPC allowed the price of new natural gas to rise to $1.42 per mcf. That price was still below the price in intrastate markets, but it did substantially close the gap between prices for new natural gas in the interstate and intrastate markets. Because the prices of old gas were controlled or determined by long-term contracts at very low prices, there was little immediate effect on the prices consumers paid for gas.

Since some interruptions are normal and expected during the winter and do not imply the existence of a shortage, it is difficult to obtain data that accurately measure the magnitude of the shortages. The Energy Information Administration has collected some data on interruptions and curtailments, but the data are provided by pipeline companies that use different criteria for deciding whether or not a given interruption of service is to be counted as a curtailment. Some include interruptions of service on interruptible contracts, and others do not. But since the same definitions continue to be used, comparisons in the aggregate figures do provide some idea of the relative severity of shortages through time. These data, beginning with the 1975-76 heating season, are reproduced in Table 10-1.

The data suggest that the shortages reached their peak in severity during the 1976-77 heating season and that they became progressively less severe, presumably as a consequence of increasing natural gas prices, both at the wellhead and to consumers in subsequent years.

Table 10-1. Curtailment of Natural Gas.

Heating Season	*Amounts Curtailed*
1975-76	1.45 tcf.
1976-77	1.99 tcf.
1978-79	1.36 tcf.
1979-80	1.21 tcf.

Source: *Monthly Energy Review*, December 1981, p. vii.

Another factor in moderating the shortages after 1978 was probably the Natural Gas Policy Act (NGPA) that became law in November 1978. The act extended price controls on prices to the intrastate market for the first time. This removed the incentive for producers to dedicate reserves to the intrastate market in which prices exceeded those in the interstate market. Against this benefit, however, one must weigh the reduced incentive for exploration caused by the ceiling now imposed on gas prices in the intrastate market. Nor does the absence of shortages in natural gas markets in the past few years mean that all distortions and inefficiencies have been eliminated from those markets any more than the absence of shortages in markets for refined products while they were subject to price controls means that those markets were free from distortions and inefficiencies.

THE NATURAL GAS POLICY ACT OF 1978

During the seventies the inefficiencies and distortions created by ceilings on natural gas prices became too costly for the political process to continue to ignore or to tolerate. Confronted with the mounting costs of its ceilings on natural gas prices, FPC abandoned its policy of freezing the price of natural gas at the wellhead, and by 1976 it had raised the ceiling on the price of newly discovered gas to $1.42 per mcf. The costs of controls on gas prices had reached such levels that Congress, notwithstanding two-to-one Democratic majorities, almost passed legislation to decontrol the price of new natural gas.

While running for President in 1976, moreover, Jimmy Carter—in a famous letter written to the Democratic governors of several gas producing states—promised to work with Congress for the decontrol of new natural gas. Barely three months after taking office, however,

Mr. Carter announced the first of several comprehensive national energy plans. In his proposal he not only failed to call for decontrol of new natural gas, but he also proposed extending controls to the intrastate market. Like most of the measures in his energy plan, Mr. Carter's proposals for maintaining and extending controls on natural gas prices evoked little enthusiasm either from Congress or the public. His proposals were unsatisfactory both to those who favored decontrol and to those who were advocating a rollback in the price of natural gas that they felt was too high already.

After over a year of negotiation and stalemate, Mr. Carter finally agreed to accept a compromise with some of those who had been supporting decontrol. NGPA embodied these compromises. It called for an immediate increase in the ceiling on the price of new natural gas to $1.75 per mcf. plus monthly inflation and incentive adjustments. It defined several different categories of gas, the prices of which were to be allowed to rise gradually toward the anticipated free market levels and were to be decontrolled either on January 1, 1985, for some categories or July 1, 1987, for others. Certain categories of new gas produced at high cost were to be decontrolled within one year, and controls were extended to the intrastate gas market.

I am now going to review briefly the principal categories of gas created by NGPA and the price provisions to which they are subject. Because of the enormous complexity of the provisions, I shall omit or simplify some.[40]

First there are two categories of new gas from wells that began production after 1976. Such gas, which comes either under section 102 or 103 of NGPA, was subject to a ceiling price of $1.75 per mcf. as of April 1977 plus a monthly inflation (and in some cases an additional) adjustment. Most of this gas was supposed to be decontrolled on January 1, 1985, but some gas from wells shallower than 5,000 feet is not scheduled to be deregulated until July 1, 1987. Certain new discoveries of offshore gas on old leases are to remain under control indefinitely.

Gas dedicated to interstate commerce before the enactment of NGPA comes under section 104. Such gas is subject to various price ceilings depending upon when it came into production and its location. As of October 1981 the ceilings ranged from $0.283 to $2.27

40. The following summary of provisions of NGPA is based on the summary found in U.S. Energy Information Administration, *Natural Gas Market* (1981).

per mcf. None of this gas will be decontrolled under the terms of NGPA.

Gas sold under existing intrastate contracts is covered by section 105. If the contract price was less than $2.078 per mcf. on November 9, 1978, the date of NGPA's enactment, the ceiling is the lower of the contract price or the ceiling under terms of section 102. If the contract price was above $2.078 per mcf. on this date, the ceiling price is the higher of the contract price plus monthly inflation adjustment or the ceiling under Section 102.

Section 106 covers the sale of gas under "rollover" contracts, that is, contracts which have expired and are renegotiated by the parties. If the gas was being sold in the interstate market, the ceiling price at the time the contract is rolled over is the higher of the ceiling on the old contract at the time of the rollover or 56¢ per mcf. as of April 1977 plus a monthly inflation adjustment. If the gas was being sold in the intrastate market, the ceiling price in the new contract is the higher of the price paid under the expired contract as of the rollover date plus a monthly inflation adjustment or $1.00 per mcf. as of April 1977 plus a monthly inflation adjustment. Old gas in the interstate market is to remain permanently under control when rolled over even after the deregulation date of new gas in the interstate market. After January 1, 1985, all gas in the intrastate market is to be decontrolled. But the rollover provisions of section 106 will continue to apply to old gas in the interstate market so long as old gas continues to be produced and sold there.

Section 107 covers certain categories of high-cost gas. Gas produced from wells 15,000 feet or deeper drilled after February 19, 1977, and gas produced from geopressured brine, coal seams, or Devonian shale were decontrolled on November 1, 1979. Gas produced from tight sands was not and will not be decontrolled, but this gas is allowed 200 percent of the section 103 ceiling price.

Stripper-well natural gas is subject to section 108. Nonassociated gas produced at an average rate less than or equal to 60 mcf. per day is subject to a ceiling price equal to $2.09 per mcf. plus monthly inflation and escalation adjustments. Stripper-well gas will not be formally decontrolled.

This brief summary cannot do justice to the Byzantine complexity of NGPA. One of the most difficult aspects of NGPA is simply to determine what section of the act is applicable to gas from a given

Figure 10-1. Maximum Ceiling Price Categories: NGPA Title I for Onshore Lower-48 Natural Gas Above 15,000 Feet.

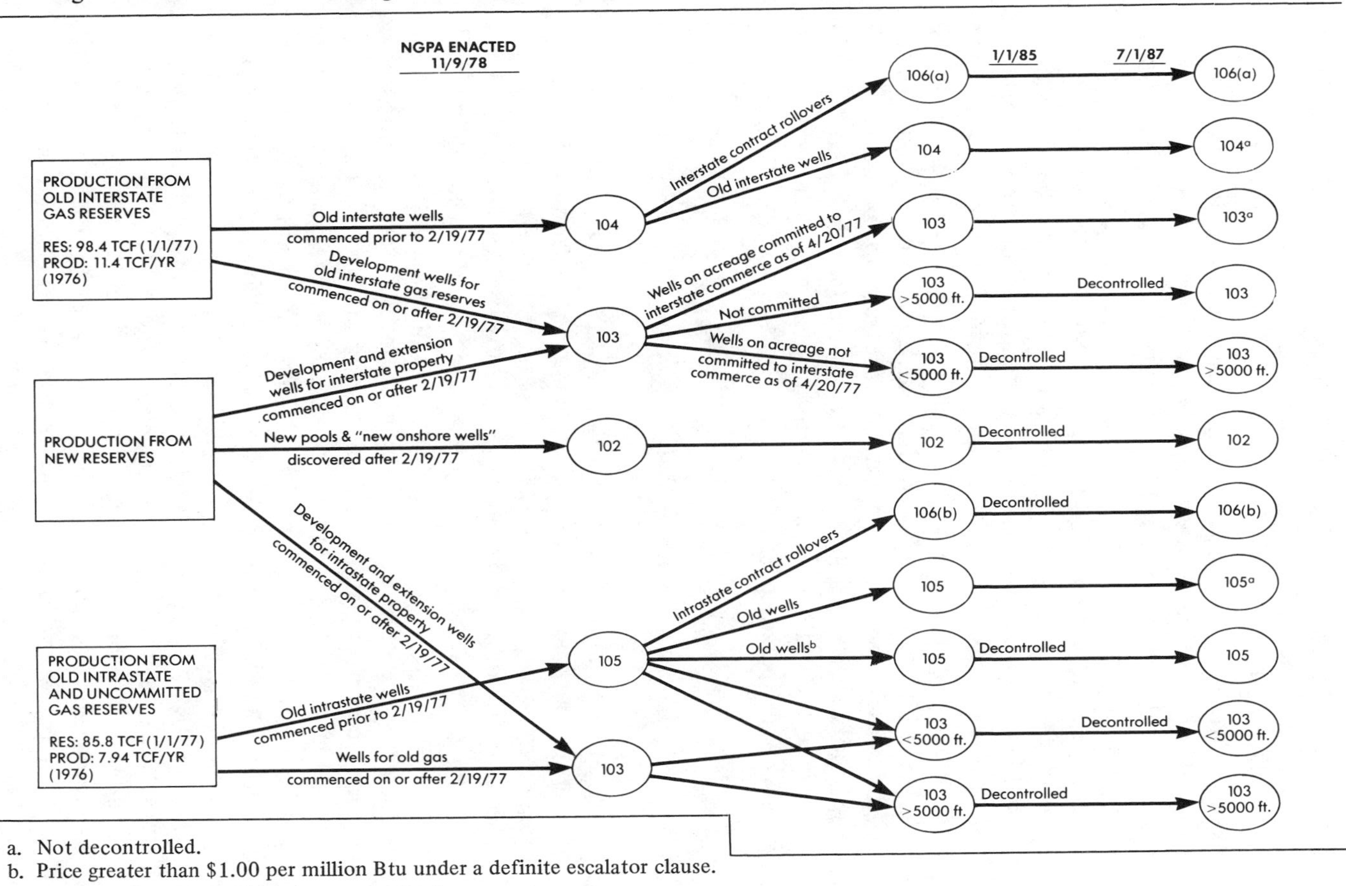

a. Not decontrolled.
b. Price greater than $1.00 per million Btu under a definite escalator clause.
Note: Stripper wells and high cost gas omitted.
Source: U.S. Energy Information Administration, *Natural Gas Market* (1981), p. 16.

well. Figure 10-1 reproduces a schematic summary of NGPA published by the Energy Information Administration.[41]

Although NGPA was a step in the direction of decontrol, it has been far from an unmixed blessing. While allowing the price of new gas sold in interstate markets to rise substantially, it has also brought gas in the intrastate markets under control. This eliminated certain distortions and reduced the drain on the interstate market, but it also reduced the incentive to find new gas. It completely decontrolled gas from high cost sources, but this created inefficient incentives to develop high-cost reserves that under normal market conditions would not have been developed before more promising, lower-cost reserves were found and developed. Moreover, NGPA did nothing to eliminate the multitier price system that had grown up as a result of the existence of long-term contracts and the regulatory distinction between old and new gas. Indeed if not further amended, NGPA would perpetuate the multitier price system until all current reserves of old gas have been exhausted.

THE MULTITIER PRICE SYSTEM

I have already given numerous examples in earlier chapters of the enormous inefficiencies created by controls that enforce different prices for similar commodities in the same market. The inefficiencies arise because they induce waste of the low-priced supplies and because resources are wasted by competition for the windfall inherent in the low-priced supply. Although attempts to maintain a low price for some share of the supply of a commodity are defended as a method of assisting consumers, my earlier analysis of the effects of such attempts in the markets for crude oil and refined products failed to support such claims on their behalf. On the contrary, I believe I have made a strong case that controls have had just the opposite effect on consumers.

A multitier price system is to some extent inevitable in a market like natural gas in which most supplies are sold under long-term contracts. The motivation for a long-term contract is, as I mentioned in Chapter 4, to avoid or to minimize the possibility that one of the parties might extort a large part of the specific capital invested by

41. U.S. Energy Information Administration, *Natural Gas Market*, p. 16.

the other party by threatening to withdraw from the transaction unless the terms of the contract are substantially improved. Stipulating the terms under which future exchanges will be made reduces the scope for such exploitative behavior. An alternative to the long-term contract is vertical integration. We find more vertical integration and fewer long-term contracts in the oil industry than in the natural gas industry. I will only refer to, without elaborating on, the hypothesis that price controls on natural gas are responsible for the fact that the natural gas industry is less vertically integrated than the oil industry.[42]

At any rate, when a long-term contract is entered into, the terms agreed upon presumably reflect some estimate of the price of the commodity in the future. If everyone expects that the price in the future will be significantly higher than the current price, the long-term price would reflect that expectation; it would have to be above the short-term price. But if conditions change unexpectedly, the prices of contracts that were entered into after conditions changed will deviate from the prices in earlier contracts. Thus even in the absence of controls, any given pipeline company is likely paying different prices for its supplies of gas depending on when the contracts were negotiated. Nevertheless, in the absence of controls such price differences would never become too great because contracts would probably contain contingency clauses allowing for price adjustments in case prices deviated substantially from the range that had been foreseen when the contracts were made.

Even under regulation the problem of multiple prices for gas was not very serious until rapidly rising oil prices lifted uncontrolled gas prices in the intrastate market (and eventually even controlled prices in the interstate market) along with them. When pipeline companies buy gas at widely disparate prices, there are two ways in which they can respond. First, in the absence of any constraints on their own pricing, they can charge a price that reflects the most costly gas they must acquire to satisfy the public's demand for gas. If pipeline companies can set prices without constraints, any low-priced gas they obtain is a windfall, pure and simple. The second possibility, which

42. For this argument see Klein, Crawford, and Alchian, "Vertical Integration," pp. 311-12, footnote 31. The authors also cite information provided by Edward Kitch that vertical integration was indeed more common in the gas industry before passage of NGA in 1938 than afterwards. But given the tremendous growth in the industry following 1938, such evidence would have to be taken with some caution.

arises because there are regulatory constraints on the rates pipeline companies charge their customers, is for pipelines to set a price that reflects the average acquisition cost of all the gas they obtain. If regulation compels such average-cost pricing, then the windfall inherent in the low-priced gas is transferred from pipelines to consumers. If so, consumers do not pay a price that reflects the marginal, or replacement, cost of those supplies. There is in effect a subsidy for the consumption of gas since one can increase the share in the windfall associated with the low-priced gas by increasing one's total consumption of gas.

It is an empirical question whether regulation in practice tends to approximate the first or the second possibility more closely. Detailed research on the way in which pipelines and local gas distributors are regulated would be required before an opinion could be rendered with confidence. I suspect that prices to consumers would be found to retain some of the windfall for the pipeline companies and to pass some of it along to consumers. How much is retained and how much is passed on is impossible to say.

However, even if we assume that pipelines do set prices on the basis of their average costs, the gains to consumers are not so large as one might at first suppose. For if there is no gas available to pipelines at uncontrolled prices, there will not be enough gas to go around (especially since consumption is being subsidized), and consumers will have to endure a shortage of gas. But if gas is available to pipelines at uncontrolled prices, the subsidy to gas consumption will induce pipelines to bid up the prices of the uncontrolled supplies to levels that will (1) attract additional supplies to meet the subsidized demand as well as (2) be high enough, when averaged with low-priced supplies, to discourage even subsidized consumption. In mid-1981 some pipelines were paying as much as $8.81 per mcf. for high-cost gas—more than four times the average price for all natural gas.[43]

Thus average-cost pricing by pipelines with access to low-priced gas must result either in shortages, if the windfall from the low-priced gas is to be preserved for the consumers, or if shortages are to be avoided, it must result in the transfer of a substantial part of the windfall to producers of uncontrolled gas. (See Appendix B.)

The passage of NGPA made available for the first time domestic sources of gas that were free of price controls to the interstate mar-

43. U.S. Energy Information Administration, *Natural Gas Market*, p. 69.

ket. Before that the only sources of supplies not subject to control were imports, mostly from Canada. But most pipelines do not have easy access to imported gas. It was only after the passage of the NGPA that the market-clearing mechanism outlined in the previous paragraph could have come into operation. Although shortages were diminishing in intensity even before NGPA was enacted, the shortages practically disappeared the year after NGPA was passed. The market-clearing mechanism embodied in NGPA may well have contributed to the elimination of shortages.

There is also evidence that as supplies have become more abundant relative to demand in recent years, pipelines have curtailed their purchases of relatively low-cost supplies in favor of purchases of high-cost gas.[44] Such behavior appears paradoxical.

Two possible explanations suggest themselves. One is that pipelines have been allowed a profit margin on their purchases of gas even though they should only be allowed a dollar-for-dollar pass-through under traditional utility regulation. If they were allowed a profit margin on such purchases when their own selling prices are calculated, it is quite conceivable that purchasing from a high-priced source is preferable to purchasing from a low-priced source because doing so permits the pipeline to sell gas at a more profitable price.[45]

A second explanation is the following. Take-or-pay requirements are usually written into long-term gas contracts for the protection of the producer, and they require the purchaser to pay for gas even if none is actually taken. If a pipeline is committed to pay for gas even if it does not take delivery, it will be more costly not to purchase from a high-cost source than from a low-cost source in a period of excess supply.[46]

44. See, for example, *The Wall Street Journal*, 11 November 1982, p. 56.

45. The principle here is similar to that which explained why refiners had an incentive to purchase supplies of gasoline for resale as a means of raising their ceiling price for gasoline. See Chapter 6, pp. 122–24.

46. Since this was written, pipeline companies faced with surplus gas that consumers were unwilling to purchase at prevailing prices have begun renegotiating the take-or-pay provisions in existing contracts. See *Oil and Gas Journal*, 15 July 1983, p. 100. With natural gas prices now in many instances above market-clearing levels, the prices for decontrolled gas have been falling drastically. See *Oil and Gas Journal*, 31 May 1982, p. 50.

DECONTROL AND THE MARKET FOR NATURAL GAS

I have been arguing in this chapter that controls on natural gas prices have caused shortages and other inefficiencies in the market for natural gas. Nevertheless, some might still argue against complete decontrol (particularly now that no shortages are being endured) on the grounds that decontrol would force consumers to pay even higher prices for gas than they are already.

However, several chapters of this book have been devoted to showing that controls on the prices of crude oil and refined products did not reduce the prices that consumers actually paid for those products. If controls on the prices of crude oil and refined products were unsuccessful in holding down the prices consumers paid for those products, is there not reason to believe that controls on the price of natural gas, especially in the absence of shortages, are also unsuccessful in reducing the price of gas to consumers? The argument that controls at the wellhead hold down prices to consumers sounds, after all, suspiciously like the one that controls on the price of crude oil hold down the price of gasoline. I am now going to consider whether the argument that controlling the price of natural gas at the wellhead holds down the price of gas to consumers is any more valid than the corresponding argument in connection with controls on the price of crude oil.

Let me begin to do so by posing the following question to those who believe that decontrol of natural gas prices at the wellhead would increase the prices consumers pay for natural gas: Are consumers now able to obtain as much natural gas as they would like at the prevailing prices? For the past few years, at least, it seems that the answer to this question is yes. The rapid increase in price over the past few years does seem to have effectively eliminated the shortages.

If this is the case, a further question presents itself: How would natural gas producers be able to sell the additional natural gas that would be forthcoming at higher prices to consumers who already can buy as much gas as they want to at current prices. To say that producers would simply be able to dictate whatever price they wanted to is no answer since dictating the price does not mean that they can also dictate how much gas consumers will buy at the higher price.

We seem to have a paradox. For unless consumers were charged lower prices, they would be unwilling to purchase the increased quantities of gas that producers would be trying to sell if they were getting paid higher prices. There are at least two ways of resolving the paradox, both of which I have already brought up earlier in the chapter. Neither approach excludes the other so that the economic forces singled out in the two approaches may be, and probably are, operating to some extent simultaneously. Let us consider them in turn.

The first approach is to take explicitly into account the fact that the price paid by consumers is not the same as the price producers receive at the wellhead. Pipeline companies and local distributors buy the gas at one price and then resell the gas to ultimate consumers at a higher price. The difference between the two prices covers their costs and may also provide them with a profit. If price controls at the wellhead have been holding down the prices paid by pipeline companies and gas distributors who have, nevertheless, apparently succeeded in raising prices to consumers to market-clearing levels, then all of the benefits of the price controls have been accruing to the pipeline companies and the gas distributors. If so, removal of the ceilings on natural gas prices would induce producers to sell more gas at higher prices while simultaneously enabling consumers to buy more gas at lower prices.

Thus our paradox is solved. Decontrol of natural gas prices at the wellhead would allow producers formerly under price controls to obtain higher prices at the wellhead by eliminating a costly subsidy to pipeline companies and to gas distributors. The cost of the subsidy is now being borne by producers subject to controls *and* by consumers.

A second approach to the paradox takes as its point of departure the multitier price system that I discussed in the previous section. The multitier price system, if it involves effective regulation on the prices that pipelines and gas distributors charge their customers, requires the averaging of the various prices paid by any pipeline or distributor in order to determine the price the pipeline or distributor may charge its own customers. As we have already seen, such average-cost pricing involves a subsidy to consumers on their purchase of gas. Now it must be clear that average-cost pricing such as this could not clear the consumer market for gas if there were not some gas that was free from an effective ceiling. Only when the price of some gas can rise enough to attract additional supplies and to increase the average-cost price sufficiently to limit the amount demanded will the market for gas clear.

This implies that the price of the decontrolled gas must rise to a level higher than it would have reached under complete decontrol because under complete decontrol, there would no longer be a subsidized demand that only a fraction of the supply could meet. Thus part of the wealth extracted from producers of old gas is dissipated by pipelines and consumers competing for supplies from the producers of decontrolled gas. Gas consumers, therefore, may pay a price that is not very far below the market-clearing price in the absence of controls. Moreover, if we assume that the supply of controlled gas would increase if controls were removed, then the additional supplies forthcoming from previously controlled gas might well prevent any price increase from occurring even with removal of the average-cost pricing subsidy to consumers. (See Appendix B.)

In the absence of any well-documented empirical evidence on the regulation of natural gas pipelines and gas distributors, it is probably safest to assume that a small part, at any rate, of the price-control subsidy is being captured by the pipelines and the gas distributors as the first approach to the paradox suggests, but some of the subsidy is also being passed through to consumers as the second approach suggests. Some indirect evidence that both of these effects are present is that both natural gas pipeline companies and producers of decontrolled high-cost gas have opposed immediate and total decontrol of all gas prices at the wellhead.[47]

Consider for example the following statement by Representative John F. Dingell, the leading congressional opponent of natural gas decontrol: "Decontrol could spell disaster for pipelines and distributors because it may make gas too expensive to market successfully."[48] The Congressman was exaggerating. Gas prices could only increase to a level at which a sufficient amount was still demanded for it to continue to be marketed. If prices rose to a level at which gas became too expensive for consumers to purchase (which is what Mr. Dingell was suggesting), the price at the wellhead would perforce

47. For evidence about opposition from producers of decontrolled gas to further decontrol, see *Oil and Gas Journal*, 1 February 1982, p. 74 and 8 February 1982, p. 64.

48. Speech of 29 January 1981, quoted in *Business Week*, 31 August 1981.

have to come down.[49] Remember it is only from the value its output has for ultimate consumers that any input or raw material (such as gas at the wellhead) has any value. But Mr. Dingell was entirely correct in observing that decontrol would force pipelines and distributors to settle for much smaller margins than they have been used to – a prospect which, understandably, does not lead them to rejoice at the prospect of decontrol.

There is another reason for supposing that the price of natural gas to consumers would necessarily rise if the price at the wellhead were decontrolled. I shall call this the Btu-equivalence theory of prices. It goes something like this. All fuels generate a certain amount of heat, which measures the amount of work they can perform. The heat content of fuels is measured by the number of Btu's per unit of fuel. Since the value of any fuel is derived from its heat content or the amount of work it can perform, the price per Btu of all fuels must be equalized. Thus if natural gas were decontrolled, its price would immediately rise to equal that of heating oil.

It requires very little sophistication to expose the fallacy on which the Btu-equivalence theory is based. Presumably the theory is not even meant to apply to the prices of all commodities, though why not is not clear. A Rembrandt, for instance, has a certain Btu content. After all we could burn it; it would generate a certain amount of heat and could perform a certain amount of work. But anyone who suggested that heating oil and Rembrandts should have the same price per Btu would be regarded with bewilderment or contempt. Yet even if we restrict the application of the Btu-equivalence theory to fuels, it is not exactly clear what reason there is for all fuels to have the same price per Btu. Fuels would only be priced equivalently if fuels were all perfectly versatile in all possible uses. But if fuels are at all specialized to certain uses (e.g., it is more costly to use residual fuel oil than it is to use gasoline to run an automobile), then the price of different fuels will depend at least as much on the significance of the operations to which they are specialized as on their Btu content.

In 1981 a dollar spent on gasoline at retail would, on average, obtain for the purchaser about 93,000 Btu's, while a dollar spent on

49. This is confirmed by the recent declines in the price of decontrolled natural gas and by the renegotiation of take-or-pay provisions in existing contracts referred to in footnote 46 above.

Table 10-2. Price of a BTU from Propane as Percentage of Same from Heating Oil.

1976	95.9%
1977	96.7%
1978	94.0%
1979	84.8%
1980	78.6%
1981	69.6%
April 1982	59.1%
1982	67.3%

Source: *Monthly Energy Review.*

residual fuel oil would have obtained 160,000 Btu's.[50] Nor is there any reason to expect that the price of natural gas per Btu would equal the price of heating oil per Btu. This was not the case in the fifties before natural gas prices came under effective control, or later in the intrastate market while natural gas prices were still uncontrolled. Nor is it true now for propane, which is the closest substitute available for natural gas. Thus both natural gas and propane have consistently been priced below heating oil in relation to their Btu equivalents.

Table 10-2 shows that between 1976 and 1981 the wholesale price per Btu of propane fluctuated between 96.7 percent and 59.1 percent of the wholesale price per Btu of heating oil. The primary explanation for this fluctuation in propane prices is that the shortages of natural gas caused by strict price controls during most of the seventies led many natural gas users to switch to propane. As a result, propane prices rose for awhile to Btu equivalence with heating oil prices. But after 1978, when higher natural gas prices finally relieved the shortage, demand for propane decreased because fewer natural gas users were unable to obtain natural gas and the price of propane relative to heating oil fell correspondingly.

This experience brings out a further point: Although there is a grain of truth in the Btu-equivalence theory, inasmuch as there may be some "normal" relationship between the prices of different fuels in respect of their Btu equivalents, the relationship is certainly not one directional. In other words, decontrolling natural gas prices could restore a "normal" relationship between the prices of natural gas and heating oil not just by increasing the price of natural gas but

50. Based on average prices for 1981 reported in the *Monthly Energy Review* and on the Btu content of gasoline and residual fuel oil reported there.

also by reducing the price of heating oil. Certainly the relaxation of controls on natural gas prices has been responsible for a substantial decrease in the price of propane relative to heating oil for which users of propane may be thankful.

Moreover, discussion of the effect of decontrol on natural gas prices normally proceeds under the assumption that decontrol of natural gas would have no impact on the world price of oil. But as we saw in Chapter 8, decontrol of domestic oil prices exerted substantial downward pressure on the world price of oil. A similar effect from the decontrol of natural gas prices is quite likely.

There are several channels through which decontrol of natural gas prices could operate to reduce the price of crude oil in world markets. I shall only mention one of them here.[51] Natural gas and crude oil are substitutes. Whenever OPEC contemplates a price change, what it must take account of is the ability of its customers to substitute other sources of energy, including U.S. natural gas. Under price controls, the price of natural gas is partially insulated from changes in the price of oil. Thus if OPEC reduced the price of oil, the price of natural gas would, at least for some considerable while, remain constant. Natural gas supply, therefore, would not fall very much if OPEC reduced its price. This implies that a price reduction by OPEC would result in a smaller increase in oil sales with controls on natural gas prices than it would without controls. Similarly, a smaller decrease in oil sales would follow an increase in oil prices with controls on gas prices than would follow without controls.

Controlling the price of natural gas, therefore, reduces the responsiveness of demand for OPEC oil to changes in the price of oil. As a consequence, the optimal price for OPEC to set for oil is higher with controls on gas prices than it would be without controls. Since there is a correlation between the world price of oil and the price of natural gas, decontrolling the price of natural gas could, before long, reduce the price of natural gas just as decontrolling the price of domestic crude oil reduced the price of domestic crude oil.

CONCLUSION

Controls on natural gas prices were a misguided enterprise from the beginning, but for a long time market conditions held gas prices in

51. The arguments are elaborated on in greater detail by Ott and Tatum, "Adverse Effects," pp. 27-46.

check so that the damage done by controls was minimal. But during the middle and late sixties an unnoticed shortage began to develop. In the early seventies this latent shortage of natural gas became all too apparent. Still no reforms were possible until the damage done by controls and shortages became too serious to be endured. Rather than abandon the controls entirely, the government stumbled around—modifying here, controlling there, and decontrolling elsewhere—until it had erected a patchwork of controls and regulation that continues to obstruct and distort normal market forces. Nevertheless, the general trend of rising prices has eliminated the shortages that had subjected large parts of the United States to potentially disastrous crises each winter during much of the seventies.

This chapter has shown that for natural gas, as for crude oil, few consumers have anything to lose from immediate and total decontrol. Only a comparatively small number of politically influential groups within the gas industry itself stand to lose from decontrol. One set of gainers from controls are the pipeline companies and the gas distributors, which (like crude oil refiners) benefit from the low prices imposed on producers. And just as they did from controls on crude oil prices, foreign oil producers gain handsomely from controls on natural gas prices that help support the world price of crude oil.

Yet despite the success of decontrol of crude oil prices, significant opposition to complete decontrol of natural gas prices continues to be expressed. Certainly insofar as the opposition stems from the effort by those who have indeed benefited from the controls on natural gas prices to retain those benefits as long as possible, such opposition is a quite understandable, if not particularly admirable, manifestation of competition in the political process. But much of the opposition to decontrol seems, at least, to stem from other causes. Some of it seems to be caused by ideological opposition to the market mechanism, while some of it seems to arise more out of ignorance about how markets work than out of ideological hostility.

While I cannot hope that the analysis I have presented here will greatly influence those who have a strong financial or ideological stake in opposing decontrol, I trust that it may dispel some of the ignorance concerning the effects of decontrol and thus contribute to more rational policymaking in the future.

PART III
PRINCIPLES

Chapter 11

ENERGY, POLITICS, AND MARKETS

All systems of preference or of restraint, therefore, being thus completely taken away, the obvious and simple system of natural liberty establishes itself of its own accord. Every man, as long as he does not violate the laws of justice, is left perfectly free to pursue his own interest his own way, and to bring both his industry and capital into competition with those of any other man, or order of men. The sovereign is completely discharged from a duty, in the attempting to perform which he must always be exposed to innumerable delusions, and for the proper performance of which no human wisdom or knowledge could ever be sufficient; the duty of superintending the industry of private people, and of directing it towards the employments most suitable to the interest of the society.

Adam Smith*

INTRODUCTION

I began this book by speaking about scarcity and competition, and in each chapter I have tried to show the many ways competition for various scarce energy resources manifests itself. My aim has been to emphasize how pervasive such competition is, for only if we recognize that markets and the political process are merely alternative mechanisms for competing to control energy and other scarce re-

*The epigraph to this chapter is taken from Adam Smith, *The Wealth of Nations*, p. 651.

sources can we begin to think sensibly about how decisions to allocate those resources ought to be made.

Unfortunately, discussions about resource-allocation decisions often seem oblivious to the fact that people use the political process to advance their interests just as diligently as they use the marketplace. If they did not ignore this fact, those who deny that a desirable outcome for society can result from the individual pursuit of selfish interests in the marketplace could not quite so easily assume that the outcomes generated by the political process would be any better. Many critics of the market – misinterpreting Adam Smith's famous remark that an individual is often "led by an invisible hand to promote an end which was no part of his intention" – seek to discredit the case in its favor by equating it with religious faith or dogma.[1] What they fail to realize, however, is that the same sort of invisible hand argument they think so naive when used to support the free market would have to be invoked in order to demonstrate the superiority of outcomes issuing from the clash of competing interests in the political process.

CONSTRAINTS AND COMPETITION

In contrast to those dismayed by the pursuit of individual self-interest in the market, I would contend that the goals individuals strive to achieve matter less than the constraints under which their competition takes place. Let me try to indicate why these constraints are so important. First, there are innumerable ways in which people could conceivably compete with each other, and many are not at all pleasant. If universal chaos and misery are to be avoided, people have to be constrained from competing to advance their own interests in ways that prevent others from advancing theirs. The constraints of private property, for example, prohibit an enormous range of competitive behavior. Thus, I may compete with a business that opens

1. This is a rhetorical tactic frequently employed by J. K. Galbraith. See for example his introduction to the third edition of *The Affluent Society*.

Even an ostensibly objective reporter for *Congressional Quarterly*, (6 March 1982, p. 525) in writing about passage of the Standby Petroleum Allocation Act despite opposition by the administration, attributed passage to the unwillingness of Congress "to buy the administration's faith that the marketplace alone can deal with a severe shortage." Why it requires greater faith to suppose that the marketplace would allocate supplies without shortages than to suppose that the political process would do so is not explained.

across the street from mine by reducing my prices or improving the quality of my products and my services, but not by burning down my competitor's premises or by threatening violence.

The notion that the free market gives people the freedom to do whatever they please is based on a misunderstanding. That kind of freedom would make the survival of any kind of social order impossible. The freedom one does have in a market economy is the freedom from infringement upon the rights to one's own person and one's property. As Smith understood the concept, freedom is characterized as much by what it prohibits as by what it permits.

Thus, to suppose that when Adam Smith spoke of an invisible hand he was simply expressing an article of religious faith or was using a rhetorical device to enlist the religious sentiments of his readers in support of the free market is incorrect. Smith was offering neither religious dogma nor propaganda for a cause; he was formulating a scientific proposition—a theorem—about the way societies work. In the terminology I have been using, his theorem can be restated as follows: Of all possible modes of competition, competition by offers of exchange (COE) is the most likely to lead to the maximization of a society's wealth and, thereby, to the improvement, of its standard of living.

PROPERTY RIGHTS AND WEALTH MAXIMIZATION

The importance of this proposition bears further comment so let me try to explain the relationship between the constraints private property imposes on our behavior and the maximization of wealth.[2] Suppose I want to use a resource. Before actually doing so, I will generally weigh the benefits I expect to derive from using it against the costs I must incur.

The constraints of private property guarantee that, under COE, these costs represent the value of the resource to other potential users. To see why, suppose all available wheat can be put to uses that are worth at least $4 a bushel. Any use of a bushel of wheat

2. The remainder of this section draws on the profound analysis of property rights and cost developed by Ronald Coase in two classic articles, "The Federal Communications Commission" and "The Problem of Social Cost."

would then entail the sacrifice of some other use worth $4. COE compels anyone who uses a bushel of wheat to bear what $4 cost because those to whom a bushel is worth at least $4 will bid up the price of a bushel to just that amount. The market price thereby keeps those who want to put wheat to uses worth less than $4 a bushel from doing what they want. Thus, COE ensures that resources are put to their most highly valued uses (that wealth is maximized) by preventing anyone from using a resource unless the resource has a greater value to that person than its cost (the most valuable foregone use).

Under no mode of competition other than COE are the values others place on a particular resource so effectively communicated to potential users. No other mode of competition forces potential users of a resource to take into account the full cost of their actions because only the constraint of private property requires potential resource users to consider the value others place on that resource. Thus, only under COE does the fact that individuals will not use resources unless they expect to derive benefits in excess of costs imply the maximization of wealth.

Perhaps another example will help to clarify this point. Suppose a cattle rancher and a wheat farmer operate on adjoining tracts of land. Under COE, the cost to the farmer of using land to grow wheat reflects the land's value to the rancher as pasture for raising cattle. If the rancher owns the land to start with, the farmer can only use the land by either buying or leasing it from the rancher, and it is easy to see that the farmer has to bear the cost of using the land to grow wheat. But even if the farmer already owns the land, the cost the farmer bears in growing wheat reflects the value of the land to the rancher since the farmer must consider what the rancher would be willing to pay in order to lease or buy the land for grazing. Thus, the benefit to the farmer from using the land will exceed the farmer's cost of using it only if the farmer values it more than the rancher. COE, in effect, causes an exchange of relevant information about the values of all potential uses of any resource, which exchange ensures that the resource is put to its most valuable use.

If, however, the constraints of private property that underlie COE were removed, there would be no mechanism to ensure that those competing for control of the land would take into account its value to other potential users. In the absence of COE, the true cost of using the land (its most valuable foregone use) would not be borne by the user, and the value of output would not be maximized.

Sometimes political competition (PC) is substituted for COE because property rights to certain kinds of resources cannot be defined or enforced or because PC is the mechanism for creating the property rights without which COE would not be possible. Property rights to the atmosphere, to clean air, or to a scenic view cannot easily be defined, enforced, or exchanged, so individuals often can only affect the ways these resources are used by resorting to PC. If, for some reason, either the cost of creating private property rights in commonly held resources decreases or the potential gains from doing so increases, the property rights may be created—as was done when the commons were enclosed in Britain and when the western territories were settled in the United States—through PC. In such cases one cannot assert that PC is less efficient than COE because the preconditions necessary for COE do not exist. But when PC is used to override or abridge existing property rights, competition for the use of resources cannot ensure that the full costs of using them are taken into account.[3]

This, of course, provides an excellent economic rationale for the provision of the Fifth Amendment to the Constitution requiring that owners be justly compensated when their property is taken for public use. If strictly abided by, the "taking clause" would impose the constraints of private property on the operation of PC and would, thereby, make PC operate more efficiently. Unfortunately, the courts have interpreted what constitutes a taking so narrowly that PC can be used to override or abridge existing property rights without compensation as long as the owner is not completely expropriated.[4]

3. However, T. L. Anderson and P. J. Hill have pointed out in a recent paper, "Privatizing the Commons," that the potential gains resulting from the establishment and assignment of property rights may be dissipated by PC in the process of establishing those rights. Dissipation occurred in the settlement of the American frontier in the last century because the various Homestead acts required resources to be expended by homesteaders in order to establish property rights over land. This compares unfavorably with the alternative private mechanisms for establishing private rights over western lands that had been used before passage of the first Homestead Act in 1862.

4. The appropriate interpretation of the "taking clause" is the subject of B. Ackerman, *Private Property and the Constitution* and is also dealt with in B. Seigan, *Economic Liberties and the Constitution.*

HOW COMPETITIVE ARE PETROLEUM MARKETS?

It might be suggested that, given the power OPEC has over the world price of oil, energy markets are not competitive and that the case for COE is simply not applicable. But the control OPEC exercises over the world price of oil is no reason for interfering with COE in domestic energy markets. Indeed, domestic controls on the prices of crude oil, refined products, and natural gas have allowed OPEC to raise prices higher than it could have in their absence. Nor was it a coincidence that the price of oil began to fall within two months after the complete decontrol of domestic crude oil prices.

The charge is also made that domestic gas and petroleum markets are not competitive either, but according to any reasonable standard, most domestic petroleum markets are not unusually concentrated. Although there may be some local markets in which, owing to special local conditions, a few firms possess some monopoly power, it cannot be plausibly argued that domestic petroleum markets in general have been monopolized.

Many of those who allege that the oil industry is not competitive recognize this and, therefore, try to support their charges by pointing to the extent of vertical integration in the industry.[5] Economic theory, however, provides no basis for the assertion that vertical integration can somehow create monopoly power that does not already exist in some upstream or downstream market. Thus, if there is no monopoly at the production level, the refining level, or the marketing level, the mere fact that firms engage in all three stages cannot create monopoly power that would not otherwise exist. Indeed, one incentive for vertical integration is to avoid exploitation by a nonintegrated firm that had a monopoly at some stage in the production process. Thus, as I explained in Chapter 4, oil companies often own their own pipelines to avoid possible exploitation by a pipeline with a local monopoly in the vicinity of the company's refinery or producing property.

5. See, for example, W. Adams, "Corporate Power and Economic Apologetics," pp. 364-66, and Senator P. Hart, "Commentary," p. 401.

WHY SOME WOULD LIKE PETROLEUM MARKETS RECONTROLLED

When domestic crude oil prices were decontrolled in 1981, proponents of continued controls were loud in their denunciations of the decision and predicted that huge increases in the prices of refined products would be the result. Since decontrol, however, crude oil and refined products prices have fallen by as much as 20 percent. Nevertheless, the remarkable success of decontrol in reducing petroleum prices generally has gone largely unacknowledged while proposals to restore controls in petroleum markets continue to be advanced.[6] At a minimum, these proposals would provide for standby presidential authority to control the prices of crude oil and petroleum products if the flow of imported oil into the United States were disrupted. Of course, given controls over prices, the proposals must also grant authority to impose mandatory allocations of crude oil and products in case of such disruptions.

If decontrol has been as much of a success as I contend and has in fact reduced the likelihood of the supply disruptions envisioned in the proposals for standby controls and allocations, why, one might ask, are proposals still being advanced to reinstitute such controls just when they would be most damaging. Four explanations come to mind for the persistence of recontrol proposals despite the manifest success of decontrol.

First, controls provide additional opportunities for various groups to engage in PC. Any group that benefited from controls in the past would almost certainly favor reimposing them once again, expecting that the controls would facilitate transfers of wealth from other less politically powerful groups to themselves.

One can get some idea of the extent to which particular groups have sought to use PC as a way of transferring wealth to themselves by analyzing the roll calls on the votes in the Senate and House of Representatives on the Standby Petroleum Allocation Act (SPAA) of 1982 and the unsuccessful vote in the Senate to override President Reagan's veto. As we saw in Chapters 5 and 6, farmers had top priority status and were, therefore, among the leading recipients of wealth transfers under the petroleum allocation programs. Not sur-

6. See Daniel Yergin, "Awaiting the Next Oil Crisis."

prisingly, agricultural organizations were among those reported to have lobbied heavily for passage of SPAA.[7]

Of the seven senators who voted against SPAA, the only Democrat was Bradley of New Jersey–one of only two states in which less than 1 percent of employed persons have agricultural jobs.[8] Of the six Republicans to vote against SPAA (Armstrong-Colorado, Humphrey-New Hampshire, Mattingly-Georgia, Nickles-Oklahoma, Percy-Illinois, and Quayle-Indiana), none was from a state in which more than 5 percent of employed persons have agricultural jobs.

The influence of agricultural interests becomes clearer in the vote of the Senate to override the veto. Democrats voted 38 to 4 to override, while Republicans voted 32 to 20 against overriding. None of the four Democrats who voted against the override (Bentsen-Texas, Boren-Oklahoma, Bradley-New Jersey, and Heflin-Alabama) was from a state where more than 5 percent of employed persons have agricultural jobs.

But what was most significant about the vote in the Senate was that of the ten senators from states (Idaho, Iowa, Nebraska, North Dakota, and South Dakota) in which more than 10 percent of those employed have agricultural jobs, all voted to override. In particular, while the eight Republicans from these states all voted to override, other senate Republicans voted 32 to 12 to uphold the veto of a Republican President.

The vote of the House of Representatives in favor of SPAA is perhaps even more revealing than the Senate vote because there are many more heavily agricultural congressional districts than heavily agricultural states. The vote in the House was 246 to 144 in favor of SPAA. Democrats voted 190 to 23 in favor, while Republicans voted 121 to 56 against. Of the 33 members of Congress from districts in which at least 10 percent of employed persons have agricultural jobs, 27 voted for SPAA and only 6 voted against. Democrats voted 12 to 1 and Republicans 15 to 5 in favor. Although Democrats from heavily agricultural districts were only slightly more disposed than other Democrats to vote for SPAA, Republicans from such districts were three times more likely than other House Repub-

7. See *Congressional Quarterly*, 27 March 1982, p. 663.

8. Data on the percentage of people employed in agricultural jobs are from M. Barone and G. Ujifusa, *Almanac of American Politics.* Roll calls of Congressional votes are from *Congressional Quarterly*, 6 March 1982 and 27 March 1982.

licans to vote for SPAA. Republicans from agricultural districts voted in favor by a 3 to 1 ratio, while other Republicans voted nearly 3 to 1 against.

A second reason for attempts to recontrol is the fear that the price increases that would result if world oil supplies were disrupted again would cause a hardship on the poor who cannot even afford to pay current prices for gasoline, heating oil, and other forms of energy—much less higher prices. The first thing to note about this argument is that although poor people might be harmed by the price increases for petroleum products, their basic problem is not higher oil prices; it is their poverty. If we want to help them, controlling oil prices is certainly not the most effective way of doing so. Moreover, controls may even create additional poverty by reducing the efficiency of the economic system. A simpler, more direct, and more effective approach would be to provide the poor with additional cash to spend in the marketplace as they see fit, or perhaps to subsidize their purchases of gasoline, heating oil, and other fuels even as we now subsidize their purchases of food.

Since the relationship between controlling oil prices and alleviating poverty is tangential at best, one suspects that much of the rhetoric about protecting the poor from the effects of higher oil prices is intended only to create a climate in which it would be possible to extract wealth from the oil industry—or to be more precise, from certain parts of it—to be redistributed through PC. It is scarcely likely that the poor will be the major beneficiaries of this process of redistribution.

The third reason I would suggest for proposals to recontrol is that the relationship between decontrol and reduced prices is still not understood. Those who continue to be enthralled by the cost-of-production fallacy I criticized in Chapter 8 are, of course, particularly susceptible to such misunderstanding and to exaggerated fears that supply disruptions would cause chaos in the absence of controls. The leading academic proponent of this view is Daniel Yergin of the Harvard Business School. Another prominent expert with the same view is James Akins, a former U.S. State Department official and former ambassador to Saudi Arabia. Both have warned of the potentially disastrous effects of disruptions in the flow of oil from the Persian Gulf. They point out that such disruptions could be occasioned by a variety of circumstances, for example, the Iran–Iraq war, renewed fighting between Israel and the Arabs, a political upheaval in a major oil producing country, or retaliation by the Arabs against the United States for being too pro-Israel.[9]

In addition to misunderstanding how market forces would minimize the impact of any possible disruption of supplies from the Persian Gulf, Yergin and Akins also fail to realize that effective cartelization of the international petroleum market has necessarily created a situation of chronic excess supply and unused production capacity. As a consequence, no threat by any oil producer to restrict its production of oil, much less to embargo any particular country, deserves to be taken seriously by sensible people.

All cartels face a critical problem: By restricting output in order to raise prices, cartel members must leave some of their productive capacity unused. Furthermore, the higher price attracts supplies from newly profitable sources. These added supplies require the cartel to restrict production even further to maintain a given price level. Thus, all cartels sooner or later are confronted with the task of how to allocate a dwindling market share to cartel members with substantial idle capacity. As a consequence, cartel members acquire an ever stronger incentive to cheat on the cartel price by offering discounts in order to increase output and reduce idle capacity.

Excess demand would be a signal that the cartel could profitably increase its price, which is not the sort of signal cartels are blind to. Price is increased until excess supply and idle capacity are no longer manageable. If, as Mr. Akins has often asserted, the Saudis are not seeking to maximize their profits but are more interested in achieving various political objectives, they would not have permitted the world price of oil to rise to levels that, except for 1979 and early 1980 during the Iranian Revolution and late 1980 after the outbreak of the Iran–Iraq war, have typically resulted in excess supplies on the world oil market since 1974.[10]

If you doubt my assertion that there will generally be an excess supply in the world oil market, just suppose that early in 1981 I had told you that during the next three years Iran and Iraq would still be waging war against each other; that Anwar Sadat would be assassinated; that the United States would shoot down two Libyan jets over the Mediterranean; that President Reagan would order American oil companies to get out of Libya; that Israel would de-

9. See D. Yergin and M. Hillenbrand, editors, *Global Insecurity*.
10. See J. E. Akins, "The Influence of Oil."

stroy an Iraqi nuclear reactor, annex the Golan Heights, and invade Lebanon, capturing and occupying the cities of Sidon and Tyre, destroying twenty Syrian antiaircraft implacements and shooting down a hundred Syrian jets; that Israel would also lay siege to Beirut, trapping Yasir Arafat and over ten thousand PLO guerillas and Syrian troops inside, and after forcing the evacuation of the PLO from Beirut, would become implicated in the massacre of hundreds of innocent Palestinian civilians by her Lebanese Christian allies; that American marines would enter Lebanon and engage in hostilities with Syrian forces; and that Saudi Arabia would cut its oil production in half. Had I told you that all of this should happen and could you have believed me, what would you have guessed would happen to the price of oil?

Despite the failure of all of these political upheavals to halt the downward trend of world oil prices following decontrol, Mr. Yergin and Mr. Akins would still have us believe that we are in imminent danger of some event that will interrupt supplies and cause prices to soar. The possibility of some such event cannot be denied, though other than a catastrophic escalation of the Iran–Iraq war, it is difficult to imagine what such an event could be. But the higher the probability of a disruption in supplies, the more dangerous are their proposals since they suggest we cope with the effects of a disruption by imposing precisely the same controls that led to gasoline shortages in the fall and winter of 1973–74 and in the spring and summer of 1979—the very controls that enabled OPEC to raise the price of oil as much as $4 a barrel above the price it could have maintained in their absence. One doesn't want to appear ungrateful, but if this is how Mr. Yergin and Mr. Akins propose to look after the interests of oil consumers, perhaps they should devote themselves to looking after the interests of OPEC instead.

There is one further reason that causes some to advocate recontrol, which is the belief that if some supply disruption cocurs, price controls and government allocation—even rationing—is fairer than allowing the allocation to be determined by the free market. It is impossible to tell how much weight this belief carries in the political process, but it is difficult to believe that it carries none. How are we otherwise to account for the marked difference in how Democrats and Republicans in Congress voted on SPAA? The Democrats, who tend to be more skeptical of the market mechanism than Republicans, voted overwhelmingly in favor of SPAA while Republicans voted less overwhelmingly against it. Certainly, the competition of interest groups for the windfalls that recontrol would create was an important factor in the final outcome, but such competition would be less likely to result in recontrol if there were not widespread misunderstanding of the effects of controls and widespread belief that controls were, in some sense, fairer than the market.

This is not the place to compare the fairness of the market with the fairness of controls. But if one views the market and the political process as alternative mechanisms through which people and groups compete to advance their own interests, then, as I have already suggested, it becomes very difficult to see any basis for supposing that the outcome of one competitive process is going to be fairer than the outcome of the other.

If the market allocates energy resources more efficiently than a mandatory allocation or rationing scheme would, and if mandatory allocations and rationing are as much driven by the pursuit of selfish interests as competition in the marketplace is, how can we explain the widely held view that rationing is a fairer means of allocating a good or resource in short supply than the market?

The explanation I would suggest is based on a certain tendency shared by most people to overrate their own honesty, intelligence, and good intentions in comparison with the honesty, intelligence, and good intentions of others. Thus, if one were to ask people randomly to rate their intelligence relative to the population, I suspect that more than 50 percent would include themselves in the smarter half. Similarly, when one asks people how they feel about a rationing system that would allocate only enough gasoline to meet everyone's legitimate needs for using it but would provide none for wasteful use, people are apt to assume that they use gasoline more wisely and judiciously than others, so that rationing would impose little hardship on them. Instead, they expect that any hardship will only be borne by others more wasteful than themselves who would thus deserve whatever hardships rationing imposed on them.

In practice, of course, no rationing system ever works out this way. Various groups are able to compete more effectively than others and use their influence to get special treatment. People quickly find out that with the amount of gasoline allocated to them they can't really do all the essential nonwasteful things they need to, and before long they begin to see other people getting more than they do; so grievances against the system rapidly accumulate.

Thus, any actual rationing system is ultimately viewed by most people who have to live with it as decidedly unfair, but they may not conclude from this that rationing is inherently unfair or that it really is not possible to allocate a commodity among countless competing uses according to any recognized principle of fairness. People are more likely to conclude that a good system is being misused—either deliberately by corrupt officials or unintentionally by incompetent ones. My conjecture, therefore, is that if one asked the public to compare a specific system of rationing with which they had direct personal experience with a free market, one would get a far different response than if one asked the public to compare rationing in the abstract with a free market.

CONCLUSION

One response to what I have been saying here might be that my argument is, in essence, that the market gives us the best of all possible worlds—that whatever outcome the market produces is efficient and should be left alone. But that would not be an accurate characterization of my position. What I have maintained is that attempts to compel others to conform to any particular view of what the world should be like—for example by controlling the prices at which crude oil and petroleum products may be exchanged or by rationing available supplies among potential users—ultimately makes things even worse. Rather than trying to impose any single fallible vision of an ideal world on others, I prefer to let them decide for themselves the terms at which they are willing to trade and the uses to which they want to devote the available supplies. In doing so, they have the opportunity to experiment in finding ways to improve the actual world we have. The market enlarges the scope for such experimentation and discovery while government control and regulation restrict it. That the market fosters a continuous process of learning and discovery, of change and adaptation, is perhaps the strongest argument one can make for not subjecting it to governmental control. To anyone who places a high value on intellectual and material progress that argument should carry great weight.

APPENDIXES

APPENDIX A

In this appendix I elaborate a bit more on the relationship between the total stock of oil and the intertemporal path followed by the price of oil. This path has to satisfy two conditions: (1) the rate of increase in the price of oil must equal the rate of interest; (2) the total demand for oil over time must equal the total stock of oil.

In Figure A–1 I have drawn two pairs of axes. Panel (a) plots price on the vertical axis and time on the horizontal axis. Panel (b) plots reserves on the vertical axis and time on the horizontal axis. If reserves at time zero equal K barrels, under given demand conditions assume that the associated price path begins at $\$X$ per barrel at time zero and thereafter rises continuously at a rate equal to the interest rate of i percent per year. The curve labeled XK in panel (a) corresponds to this price path. In panel (b), I draw a curve showing what happens to reserves over time if price path XK is followed. This curve, labeled KK, may take a variety of shapes depending on what demand is like. The curve I have drawn corresponds to a special assumption about a constant rate of growth of demand and a constant elasticity of demand[1] along the demand curve and over time.[2]

1. See Williams, "Running Out" or Dasgupta and Heal, *Economic Theory*, for derivations.

2. Elasticity of demand represents the ratio of the percentage change in quantity to the corresponding percentage change in price. It is a measure of the responsiveness of demand to price changes. The more responsive demand is, the greater the elasticity.

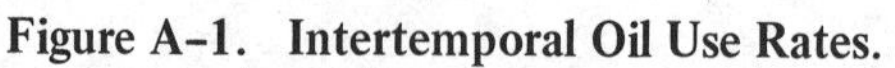

Figure A-1. Intertemporal Oil Use Rates.

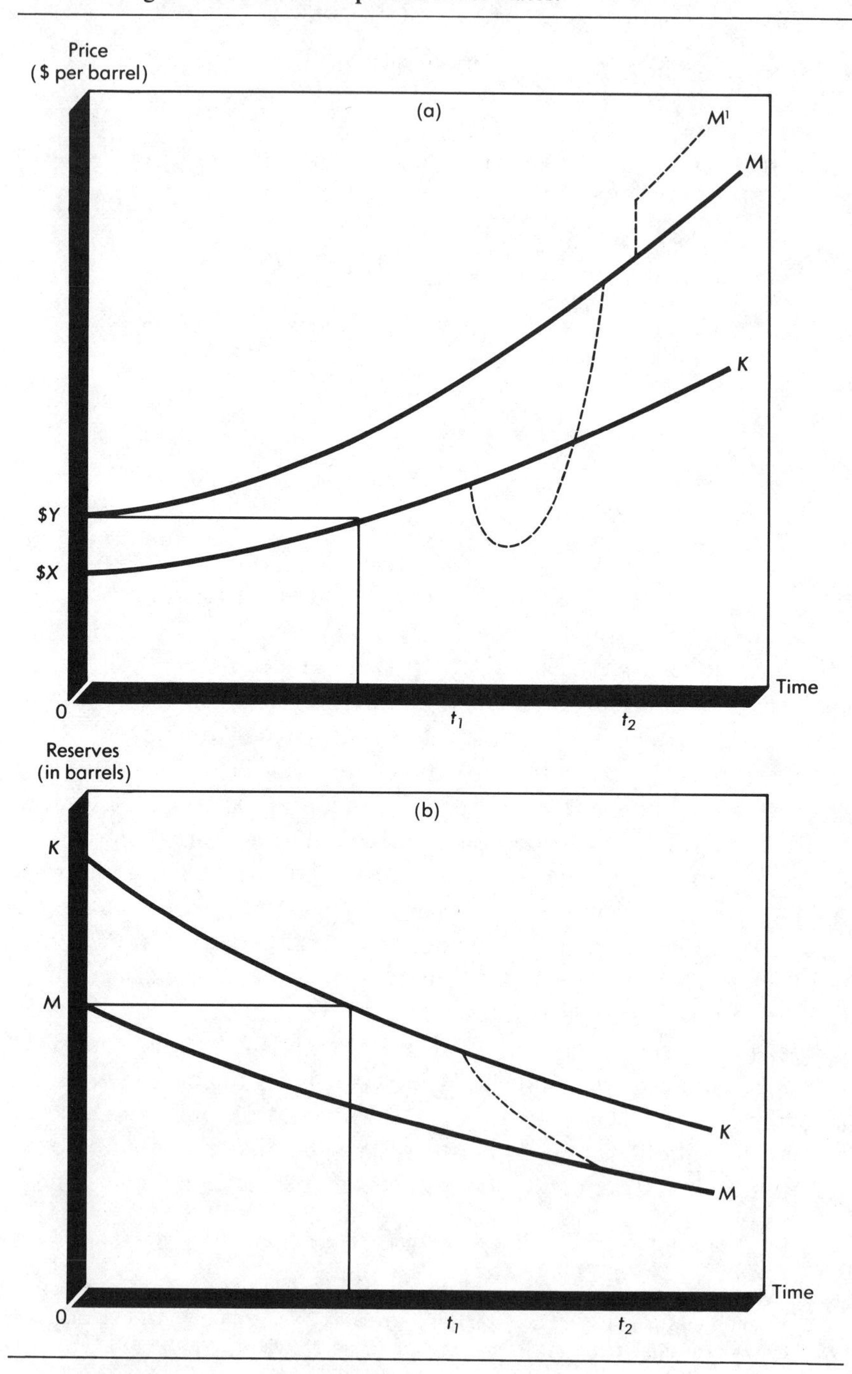

In such a case, it can be shown that the two conditions would be satisfied if during any period consumption were equal to a certain constant fraction of reserves at the beginning of the period.[3]

If reserves were initially equal to *M* barrels instead of *K* barrels and *M* is less than *K*, the initial price would have to be say $*Y* per barrel instead of $*X* per barrel, and *Y* would be greater than *X*. Thus in panel (a) I have drawn curve *YM* corresponding to the price path that would be followed under identical demand conditions if the initial supply were *M* barrels instead of *K* barrels. In panel (b) the curve *MM* indicates what happens to reserves over time if the price path *YM* is followed. Note also that on *XK* when the price equals $*Y* per barrel, the corresponding amount of reserves on curve *KK* in panel (b) is *M*.

One can readily see that if, instead of following path *XK*, the price were, at time t_1, to fall below the path because extraction was too rapid, reserves would fall below the curve *KK* at that moment. With no upward adjustment of price, extraction and consumption would ultimately exhaust reserves. If price continued below *XK*, reserves would continue below *KK*, and exhaustion would eventually occur. But well before exhaustion occurred, the prospect of higher future prices caused by dwindling reserves would provide an incentive for increased investment in reserves and a slower rate of extraction than that corresponding to *KK*. The result would be an increase in price that shifted the actual price path above *XK*. Insofar as the later price increases above *XK* are foreseen at time t_1, speculators would increase their withholding of oil when prices fell below the path and thus limit the ultimate deviation from the path.

Now suppose there is, at time t_2, an overinvestment in oil that causes the actual price to exceed the price path labeled *YM*. It is at least possible that there would be no readjustment to a lower price path and that reserves would remain above *MM*. The price could continue to appreciate, and owners of oil could continue to invest in oil on the aasumption that prices would continue to rise at a rate equal to the interest rate. Their expectations would not necessarily be falsi-

3. In a case of zero growth in demand over time and a uniform elasticity of demand equal to one, the fraction would equal the rate of interest. So if the rate of interest equaled 5 percent per year, 5 percent of the stock at the beginning of each year would be consumed during each year. This implies that consumption each year would fall by 5 percent as is required, when the elasticity of demand equals one, by the increase in price of 5 percent per year.

fied as they necessarily would in the previous case. Thus, some available oil might never be consumed. That would clearly be a waste.[4]

The market forces counteracting underconsumption do not seem as powerful as those counteracting overconsumption, since the plans on which underconsumption would be carried out are not inconsistent with the amount of oil available as are the plans based on overconsumption. Nevertheless, the greater reserves are relative to consumption, the higher the probability of an ultimate price reduction or deceleration in the appreciation of oil. All owners would fear that other owners might try to sell off the excess stocks rather than continue to hold them. To avoid taking capital losses, owners would extract their excess oil reserves and thereby move back in the direction of the lowest sustainable price path consistent with existing reserves.

4. Dasgupta and Heal, *Economic Theory*, pp. 161-63.

APPENDIX B

This appendix will show somewhat more rigorously than in the body of Chapter 10 how average-cost pricing (ACP) of natural gas by pipelines under a two-tier or multitier price system at the wellhead (1) subsidizes consumption, (2) subsidizes producers of uncontrolled supplies, and (3) leads to market clearing if one source of supply to pipelines is free from controls.[1]

Suppose that there are two sources of gas supply to pipelines, old and new. Old gas is available up to the maximum rate of output at the ceiling price. New gas is available in increasing quantities as its price rises, starting from the ceiling price on old gas. These supply conditions are represented in Figure B–1 by supply curves S_o and S_n. The total supply of gas is represented by S_t, which corresponds to the horizontal summation of the supply curves of old and new gas. Consumer demand for gas is represented by the negatively sloped straight line labeled *D*.

To simplify the analysis, I am going to assume that pipelines operate costlessly except for buying supplies at the wellhead. Thus the price to the consumer is the average price paid by the pipeline. Now

1. The model used here is based on that employed by Reuben Kessel to explain the behavior of milk when the price received is an average of prices in different markets. The result in Kessel's model is a surplus of milk. See Kessel, "Regulation of Milk Markets," pp. 58–60. The relevance of the model for the natural gas market was suggested by Kitch, "Regulation for Natural Gas," p. 279, foonote 123.

Figure B-1. Two-Price System for Gas Under Partial Decontrol.

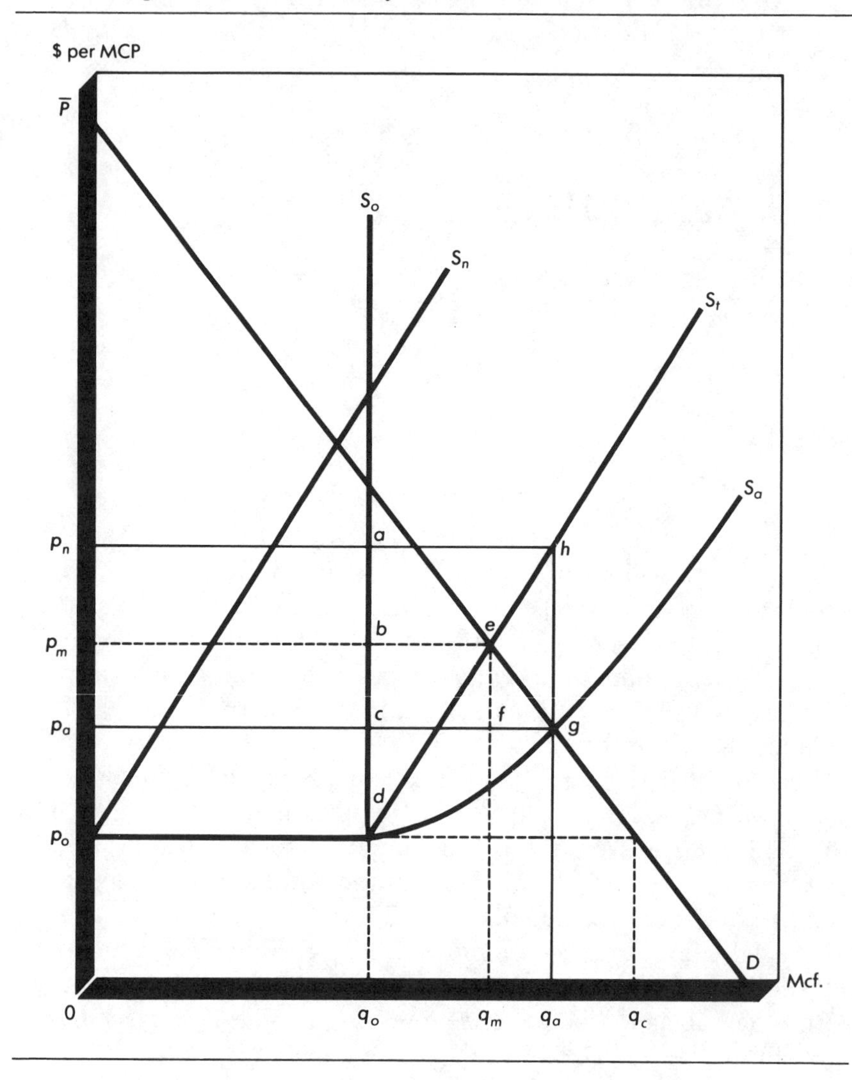

it can be seen from the diagram that if the ceiling on the price of old gas, p_o, were applied to all gas, no new gas would be forthcoming, and the amount of old gas supplied, q_o, would be less than the amount demanded at that price, q_c. The difference between q_c and q_o represents the unsatisfied demand or the shortage attributable to the price ceiling.

Let us now see what would happen if new gas were decontrolled. If the price charged consumers corresponded to the marginal cost, the price paid would equal whatever price pipelines had to pay for new gas. The market equilibrium would thus correspond to the intersection of S_t and D. Consumers would pay price p_m and would consume quantity q_m of gas. Pipelines would earn a profit equal to rectangle $p_o p_m bd$. The profit is derived from the windfall on old gas sold to consumers at the same price, p_m, pipelines pay for new gas but for which they paid only p_o.

But if pipelines must set their price to consumers equal to average cost, the price consumers pay will be less than the price received by producers of new gas. The price consumers pay under ACP is represented by the curve S_a, which begins at point b and slopes upward but lies below S_t. Under ACP, market equilibrium occurs at the intersection of D and S_a. Consumers pay price p_a, which is the weighted (by quantity) average of price p_o paid for old gas and price p_n paid for new gas. Price p_n is clearly greater than price p_m under marginal-cost pricing (MCP). Price p_a is, however, still below price p_m. The difference between p_m and p_a can be thought of as the subsidy per unit of consumption. Similarly, the difference between p_n and p_m can be thought of as the per-unit subsidy to the production of new gas. But since the total subsidy cannot exceed the difference between the market value and the controlled value of the old gas, the per-unit subsidy on consumption and the per-unit subsidy on the production of new gas will depend on the elasticity of the supply of new gas and the elasticity of demand for gas.

One way of analyzing the effects of ACP under a two-price system is to compare consumers' surplus,[2] the profits of pipelines, and the profits of producers of new gas under MCP and ACP. We have the following relationships:

revenue to pipelines under MCP $= Op_m eq_m$

cost to pipelines under MCP $= Op_o dq_o + q_o beq_m$

2. Consumers' surplus is the difference between the maximum amount consumers would have been willing to pay for what they bought (which is approximated by adding the area beneath the demand curve up to the quantity purchased) and what they actually paid (which is given by the rectangle formed by multiplying the price paid times the quantity purchased). The phenomenon arises because consumers purchasing at a flat price pay only as much per unit as the value of the marginal unit purchased. Presumably inframarginal units had a greater value. This is what a downward sloping demand curve implies. For a further explanation of the concept of consumers' surplus, see any introductory economics textbook.

profit to pipelines under MCP	$= p_o p_m bd$
consumers' surplus under MCP	$= p_m \overline{p}e$
revenue to new producers under MCP	$= q_o beq_m$
cost to new gas producers under MCP	$= q_o deq_m$
profit to new gas producers under MCP	$= dbe$
revenue to pipelines under ACP	$= Op_a gq_a$
cost to pipelines under ACP	$= Op_o dq_o + q_o ahq_a$
	$= Op_a gq_a$
consumer's surplus under ACP	$= p_a \overline{p}g$
revenue to new gas producers under ACP	$= q_o ahq_a$
cost to new gas producers under ACP	$= q_o dhq_a$
profit to new gas producers under ACP	$= dah$
loss by pipelines under ACP instead of MCP	$= p_o p_m bd$
gain by consumers under ACP instead of MCP	$= p_a p_m eg$
gain by new gas producers under ACP instead of MCP	$= bahe$
net loss under ACP instead of MCP	$= ehg$

The gain that pipelines would have enjoyed under MCP is dissipated by the competition of new gas producers and consumers under ACP. New gas producers are encouraged to produce more gas by the subsidy on gas production, but the additional gas they produce, $q_m q_a$, is more costly than its value to consumers. The excess of cost over value is represented by the vertical distance between s_t and D.[3] Consumers are encouraged to increase their consumption of gas, and the subsidy enables them to purchase gas that would otherwise be more expensive to them than it is worth. Thus part of the loss to pipelines is transferred directly to consumers in the form of a lower price on the amount of old gas that had been purchased, q_o. This transfer from pipelines corresponds to $p_a p_m bc$.

Consumers also pay a lower price for new gas they had been purchasing, $q_o q_m$. This transfer from pipelines corresponds to *bcef*.

3. It is the cost of producing additional units of gas that determines position of S_t. The vertical distance between the horizontal axis and S_t represents the marginal cost of production at the corresponding quantity.

Consumers also gain *feg*, which is the value of their additional gas purchases, $q_m q_a$. New gas producers increase their production and obtain an increase in profit equal to *bahe.* Gas pipelines, in addition to the direct transfer of $p_a p_m bc$ to consumers, also lose $p_o p_a cd$. But this amount is necessarily equal to *cahg* since, under ACP, total revenue is equal to total cost for pipelines. Of *cahg, cbeg* goes to consumers as a further transfer from pipelines and an increase in consumer surplus, and *bahe* goes to new gas producers as increased profit. But *ehg* is a pure loss to society since there is no corresponding gain to offset the loss to pipelines. It arises because the extra production of high cost gas is not worth as much to consumers as it costs to produce.

Figure B-2 shows the situation under complete decontrol and abandonment of the two-price system. It assumes that there is some potential increase in the output of old gas that would be possible if a higher price were allowed. Thus beyond q_o, s_o is shown as shifting to the right. There is also reason to suppose that s_t would shift to the right as well since elimination of price controls would signal that higher prices in the future would accrue to producers and not be extracted by price controls. This would increase the incentive to explore for and develop reserves in the present so the supply curve should shift to the right. Thus s_t must also shift to the right under decontrol. At the very least, this reduces the gap between p_m and p_a, and it is possible that p_m could even fall below p_a.

Figure B-2. Single-Price System for Gas Under Full Decontrol.

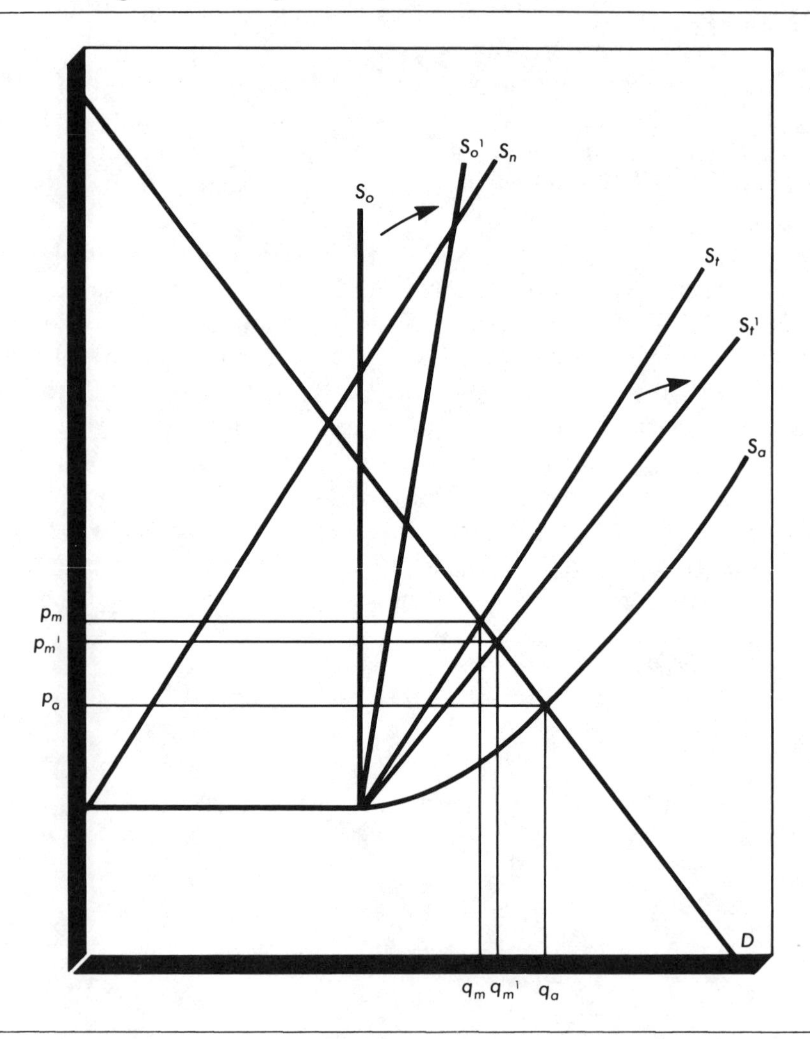

BIBLIOGRAPHY

Ackerman, Bruce J. *Private Property and the Constitution.* New Haven, Conn.: Yale University Press, 1978.

Adams, Walter. "Corporate Power and Economic Apologetics: A Public Policy Perspective." In Harvey J. Golschmid, H. Michael Mann, J. Fred Weston, eds., *Industrial Concentration: The New Learning.* Boston: Little Brown, 1974, pp. 360-77.

Adelman, Morris A. *The World Petroleum Market.* Baltimore, Md.: Johns Hopkins University Press, 1972.

Akins, James E. "The Influence of Politics on Oil Pricing and Production Policies." *Arab Oil and Gas* 10 (16 October 1981): 19-28.

Alchian, Armen A. *Economic Forces at Work.* Indianapolis: Liberty Press, 1977.

____, and William R. Allen. *University Economics.* Belmont, Calif.: Wordsworth Publishing Company, 1972.

____, and Harold Demsetz. "The Property Rights Paradigm." *Journal of Economic History* (March 1973): 16-27.

Allvine, Fred C., and James M. Patterson. *Highway Robbery: An Analysis of the Gasoline Crisis.* Bloomington, Ind.: University of Indiana Press, 1974.

American Petroleum Institute. *Market Shares and Individual Company Data for U.S. Energy Markets, 1950-1980.* Discussion Paper #014R, 9 October 1981.

Anderson, Terry L., and Peter J. Hill. "Privatizing the Commons: An Improvement." *Southern Economic Journal* 50 (October 1983): 438-50.

Anderson, Terry L., ed. *Water Rights: Scarce Resource Allocation, Bureaucracy, and the Environment.* San Francisco: Pacific Institute for Public Policy Research, 1983.

Arrow, Kenneth J., and Joseph P. Kalt. *Petroleum Price Regulation: Should We Decontrol?* Washington, D.C.: American Enterprise Institute, 1979.

Baird, Charles W. *Prices and Markets*, 2d ed. St. Paul, Minn.: West, 1981.

Barnett, Harold J., and Chandler Morse. *Scarcity and Growth: The Economics of Natural Resource Availability.* Baltimore, Md.: Johns Hopkins University Press, 1963.

Barone, Michael, and Grant Ujifuaa. *Almanac of American Politics 1982.* Washington, D.C.: M. Barone and Company, 1982.

Barzel, Yoram, and Christopher D. Hall. *The Political Economy of the Oil Import Quota.* Stanford, Calif.: Hoover Institution Press, 1977.

Blair, John M. *The Control of Oil.* New York: Random House, 1977.

Bohi, Douglas R., and Milton Russell. *Limiting Oil Imports.* Baltimore, Md.: Johns Hopkins University Press, 1978.

Bork, Robert H. "Vertical Integration and the Sherman Act: The Legal History of an Economic Misconception." *University of Chicago Law Review* (Autumn 1954): 157-201.

Breyer, Stephen G., and Paul W. MacAvoy. "Regulating Natural Gas Producers." In Robert J. Kalter and William A. Vogley, eds., *Energy Supply and Government Policy.* Ithaca, N.Y.: Cornell University Press, 1976, pp. 161-92.

Brock, Jonathan, and Roger Winsby. "Removing Controls: The Policy of Selective Decontrol." In U.S. Treasury Department, *Historical Working Papers on the Economic Stabilization Program: August 1971 to April 30, 1978.* Washington, D.C.: Government Printing Office, 1974, pp. 859-948.

Bupp, I. C., and Frank Schuller. "Natural Gas: How to Slice a Shrinking Pie." In Robert Stobaugh and Daniel Yergin, eds., *Energy Future.* New York: Random House, 1979.

Burke, Edmund. *Tract on the Popery Laws* in *The Works of The Right Honorable Edmund Burke*, 6th ed., vol. 5. Boston: Little, Brown and Company, 1880.

Cabinet Task Force on Oil Import Control. *The Oil Import Question: A Report on the Relationship of Oil Imports to the National Security.* Washington, D.C.: Government Printing Office, February 1970.

Cannan, Edwin. *An Economist's Protest.* London: P. S. King and Son, 1927.

Coase, Ronald H. "The Federal Communications Commission." *Journal of Law and Economics* 3 (October 1959): 1-40.

____. "The Problem of Social Cost." *Journal of Law and Economics* 3 (October 1960): 1-44.

Cook, Fred J. "The Great Daisy Chain Scandal." *The Nation* 229 (July 7, 1979): 1-6.

Daly, Herman. *Steady State Economics.* San Francisco: W. H. Freeman, 1977.

Dam, Kenneth W. "Implementation of Import Quotas: The Case of Oil." *Journal of Law and Economics* (April 1971): 1-60.

Dasgupta, P. S., and G. M. Heal. *Economic Theory and Exhaustible Resources.* Cambridge, Mass.: Cambridge University Press, 1979.

Deacon, Robert T., and M. Bruce Johnson, eds. *Forestlands: Public and Private.* San Francisco: Pacific Institute for Public Policy Research, forthcoming.

Demsetz, Harold. "Toward a Theory of Property Rights." *American Economic Review* (May 1967): 347-59.

Douglas, Paul. "The Case for the Consumer of Natural Gas." *Georgetown Law Journal* (June 1956): 566-606.

Fleming, Harold M. *Gasoline Prices and Competition.* New York: Meredith, 1966.

Frey, John W., and H. Chandler Ide. *A History of the Petroleum Administration for War, 1941-1945.* Washington, D.C.: Government Printing Office, 1946.

Friedman, Milton. "Real and Pseudo Gold Standards." *Journal of Law and Economics* (October 1961): 1-11.

Galbraith, John Kenneth. *The Affluent Society.* 3d ed. Boston: Houghton Mifflin, 1976.

Georgescu-Roegen, Nicholas. *The Entropy Law and the Economic Process.* Cambridge, Mass.: Harvard University Press, 1971.

Gerwig, Robert. "Natural Gas Production: A Study of Costs of Regulation." *Journal of Law and Economics* (October 1962): 69-92.

Glasner, David. "The Return of Gasoline Price Wars." *Policy Report* (August 1981): 1-3.

____ . "Relative Prices and Price Levels." *New Leader* (April 5, 1982): 10-12.

Gordon, H. Scott. "Economics and the Conservation Question." *Journal of Law and Economics* (October 1958): 110-21.

Griffin, James M., and Henry B. Steele. *Energy Economics and Policy.* New York: Academic Press, 1980.

Hart, Phillip A. "Commentary." In Harvey J. Goldschmid, H. Michael Mann, and J. Fred Weston, eds., *Industrial Concentration: The New Learning.* Boston: Little Brown, 1974, pp. 400-403.

Harvey, Scott, and Calvin T. Roush, Jr. *Petroleum Production Price Regulations: Output Efficiency and Competitive Effects.* Staff Report of the Bureau of Economics to the Federal Trade Commission. Washington, D.C.: Federal Trade Commission, February 1981.

____ and ____ . "Petroleum Product Price Regulations: Output, Efficiency, and Competitive Effects." In *Carnegie-Rochester Conference Series on Public Policy* (Spring 1981): 109-45.

Hayek, F. A. "The Use of Knowledge in Society." *American Economic Review* (March 1945): 77-91.

____. *Individualism and Economic Order.* Chicago: University of Chicago Press, 1948.

____. *The Constitution of Liberty.* Chicago: University of Chicago Press, 1960.

____. *New Studies in Philosophy, Politics, Economics, and the History of Ideas.* Chicago: University of Chicago Press, 1978.

____. *The Fatal Conceit.* Chicago: University of Chicago Press, forthcoming 1985.

Hogarty, Thomas F. "The Origin and Evolution of Gasoline Marketing." Research Study #022. Washington, D.C.: American Petroleum Institute, 1981.

Hotelling, Harold. "The Economics of Exhaustible Resources." *Journal of Political Economy* (March 1931): 137-75.

Jevons, William Stanley. *The Coal Question.* London: Macmillan, 1865.

Johnson, Harry G. "Political Economy Aspects of International Monetary Reform." *Journal of International Economics* (September 1972): 401-23.

____. *On Economics and Society.* Chicago: University of Chicago Press, 1977.

Johnson, William A. "The Impact of Price Controls on the Oil Industry: How to Worsen an Energy Crisis." In Gary Eppen, ed., *Energy: The Policy Issues.* Chicago: University of Chicago Press, 1975, pp. 99-121.

Joskow, Paul, and Robert Pindyck. "Those Subsidized Energy Schemes." *Wall Street Journal*, 2 July 1979, p. 10.

Kalt, Joseph P. *The Economics and Politics of Oil Price Regulations: Federal Policy in the Post-Embargo Era.* Cambridge: MIT Press, 1981.

Kardatzke, Nyle B. "Intercity Differences in Retail Gasoline Price Fluctuations." Ph.D. dissertation, University of California at Los Angeles, 1978.

Kay, John A., and James A. Mirrlees. "The Desirability of Natural Resource Depletion." In D. W. Pearce and J. Rose, eds., *The Economics of Natural Resource Depletion.* 1975, pp. 140-76.

Kessel, Reuben A. "Economic Effects of Federal Regulation of Milk Markets." *Journal of Law and Economics* (October 1967): 51-78.

Kitch, Edmund W. "Regulation of the Field Market for Natural Gas by the Federal Power Commission." *Journal of Law and Economics* (October 1968): 243-80.

Klein, Benjamin, Robert G. Crawford, and Armen A. Alchian. "Vertical Integration, Appropriable Rents and the Competitive Contracting Process," *Journal of Law and Economics* (October 1978): 297-326.

____, and Keith B. Leffler. "The Role of Market Forces in Assuring Contractual Performance." *Journal of Political Economy* (August 1981): 615-41.

Knight, Frank H. *Risk, Uncertainty, and Profit.* Chicago: University of Chicago Press, 1980.

Lane, William C., Jr. *The Mandatory Petroleum Price and Allocation Regulations: A History and Analysis.* Report to the American Petroleum Institute, Washington, D.C., 1981.

Libecap, Gary D. *Locking Up the Range: Federal Land Controls and Grazing.* San Francisco: Pacific Institute for Public Policy Research, 1981.

MacAvoy, Paul W. *Price Formation in Natural Gas Fields.* New Haven, Conn.: Yale University Press, 1962.

McDonald, Stephen L. *Petroleum Conservation in the United States.* Austin, Tex.: University of Texas Press, 1970.

McGee, John S. "Predatory Price Cutting: The Standard Oil (N.J.) Case." *Journal of Law and Economics* (October 1958): 137-70.

Mancke, Richard. *Squeaking By: U.S. Energy Policy Since the Embargo.* New York: Columbia University Press, 1976.

Meadows, Donella H., Dennis L. Meadow, Jorgen Randers, and William H. Behren III. *The Limits to Growth.* New York: Potomac Association, 1972.

Murphy, Frederic H.; Richard P. O'Neil; and Mark Rodekohr. "An Overview of Natural Gas Markets." *Monthly Energy Review* (December 1981): i-viii.

O'Neill, Richard P. "The Interstate and Intrastate Natural Gas Markets." *Monthly Energy Review* (January 1982): i-ix.

Ott, Mack, and John A. Tatum. "Are There Adverse Inflation Effects Associated with Natural Gas Decontrol?" *Contemporary Policy Issues*, supplement to *Economic Inquiry* (October 1982).

Owens, Charles R. "History of Petroleum Price Controls." In U.S. Department of Treasury, *Historical Working Papers on the Economic Stabilization Program: August 15, 1971, to April 30, 1974.* Washington, D.C.: Government Printing Office, pp. 1223-1340.

Phelps, Charles E., and Rodney T. Smith. *Petroleum Regulation: The False Dilemma of Decontrol.* Santa Monica, Calif.: Rand Corporation, January 1977.

Popper, Karl R. *The Poverty of Historicism.* Princeton, N.J.: Princeton University Press, 1957.

____. *The Open Universe: An Argument for Indeterminism.* Totowa, N.J.: Rowman and Littlefield, 1982.

Rawls, John. *A Theory of Justice.* Cambridge, Mass.: Harvard University Press, 1971.

Rifkin, Jeremy, with Ted Howard. *Entropy: A New World View.* New York: Viking Penguin, 1979.

Robertson, Sir Dennis H. *Economic Commentaries.* London: Staples Press, 1954.

Sampson, Anthony. *The Seven Sisters.* New York: Viking Press, 1975.

Schurr, Sam H.; Joel Darmstadter; Harry Perry; William Ramsey; and Milton Russell. *Energy in America's Future: The Choices Before Us.* Baltimore, Md.: Johns Hopkins University Press, 1979.

Scott, Anthony. *Natural Resources: The Economics of Conservation.* Toronto: Toronto University Press, 1955.

Siegan, Bernard. *Economic Liberties and the Constitution.* Chicago: University of Chicago Press, 1982.

Simon, Julian A. *The Ultimate Resource.* Princeton, N.J.: Princeton University Press, 1981.

Smith, Adam. *The Wealth of Nations.* New York: Modern Library, 1937.

Smith, Rodney T. "In Search of the 'Just' U.S. Oil Policy: A Review of Arrow and Kalt and More," *Journal of Business* (January 1981): 87-116.

Stigler, George J. *The Citizen and the State.* Chicago: University of Chicago Press, 1975.

____. *The Economist as Preacher and Other Essays.* Chicago: University of Chicago Press, 1982.

Stroup, Richard L., and John A. Baden. *Natural Resources: Bureaucratic Myths and Environmental Management.* San Francisco: Pacific Institute for Public Policy Research, 1983.

Thompson, Earl A. "An Economic Basis for the 'National Defense' Argument for Aiding Certain Industries." *Journal of Political Economy* (February 1979): 1-36.

U.S. Department of Energy. *Final Report to the President on Oil Supply Shortages During 1979.* Washington, D.C.: U.S. Department of Energy, July 1980.

U.S. Department of Energy. *The State of Competition in Gasoline Marketing.* Washington, D.C.: U.S. Department of Energy, December 1981.

U.S. Department of Justice. *Report of the Department of Justice to the President Concerning the Gasoline Shortage of 1979.* Washington, D.C.: U.S. Department of Justice, July 1980.

U.S. Energy Information Administration. *The Current State of the Natural Gas Market: An Analysis of the Natural Gas Policy Act and Several Alternatives,* pt. I. Washington, D.C.: Government Printing Office, December 1981.

U.S. President (James E. Carter). *The National Energy Plan.* Washington, D.C.: Government Printing Office, April 1977.

Wanniski, Jude. *The Way the World Works.* New York: Basic Books, 1978.

Wicksteed, Phillip H. *The Common Sense of Political Economy.* New York: Augustus M. Kelly, 1967.

Williams, Stephen F. "Running Out: The Problem of Exhaustible Resources." *Journal of Legal Studies* (January 1978): 165-99.

Yergin, Daniel. "Awaiting the Next Oil Crisis." *New York Times Magazine,* 11 July 1982, pp. 18, 22, 24, 52, 56, 58.

____, and Martin Hillenbrand, eds. *Global Insecurity: A Strategy for Energy and Economic Renewal.* Boston: Houghton Mifflin, 1982.

INDEX

ABOUT THE AUTHOR

David Glasner is Senior Analyst with National Economic Research Associates. Previously he was Assistant Professor of Economics at Marquette University (1976–83) and Visiting Assistant Professor of Economics at New York University (1983); and has taught at the University of California at Los Angeles (1970–73), California State University at Long Beach (1973 and 1975), California State University at Northridge (1974), and the University of California at Riverside (1976).

Dr. Glasner has focused his research interests on industrial organization, the economic analysis of law, macroeconomics, monetary theory, and the history of economic thought, and is the author of "The Effect of Rate Regulation on Automobile Insurance Premiums." He has written numerous articles for scholarly and popular publications, including the *History of Economics Society Bulletin, Intercollegiate Review, International Review of Law and Economics, National Review, New Leader*, and *Policy Report.*